Undergraduate Lecture Notes in Physics

T0255052

For further volumes:
http://www.springer.com/series/8917

Undergraduate Lecture Notes in Physics (ULNP) publishes authoritative texts covering topics throughout pure and applied physics. Each title in the series is suitable as a basis for undergraduate instruction, typically containing practice problems, worked examples, chapter summaries, and suggestions for further reading.

ULNP titles must provide at least one of the following:

- An exceptionally clear and concise treatment of a standard undergraduate subject.
- A solid undergraduate-level introduction to a graduate, advanced, or nonstandard subject.
- A novel perspective or an unusual approach to teaching a subject.

ULNP especially encourages new, original, and idiosyncratic approaches to physics teaching at the undergraduate level.

The purpose of ULNP is to provide intriguing, absorbing books that will continue to be the readers preferred reference throughout their academic career.

Series Editors

Neil Ashby
Professor, Professor Emeritus, University of Colorado Boulder, CO, USA

William Brantley
Professor, Furman University, Greenville, SC, USA

Michael Fowler
Professor, University of Virginia, Charlottesville, VA, USA

Michael Inglis
Professor, SUNY Suffolk County Community College, Selden, NY, USA

Elena Sassi
Professor, University of Naples Federico II, Naples, Italy

Helmy Sherif
Professor Emeritus, University of Alberta, Edmonton, AB, Canada

Ilya L. Shapiro • Guilherme de Berredo-Peixoto

Lecture Notes on Newtonian Mechanics

Lessons from Modern Concepts

 Springer

Ilya L. Shapiro
Departamento de Fisica – ICE
Universidade Federal de Juiz de Fora
Juiz de Fora, Brazil

Tomsk State Pedagogical University
Tomsk, Russia

Guilherme de Berredo-Peixoto
Departamento de Fisica – ICE
Universidade Federal de Juiz de Fora
Juiz de Fora, Brazil

ISSN 2192-4791   ISSN 2192-4805 (electronic)
ISBN 978-1-4614-7824-9   ISBN 978-1-4614-7825-6 (eBook)
DOI 10.1007/978-1-4614-7825-6
Springer New York Heidelberg Dordrecht London

Library of Congress Control Number: 2013940515

Printed on acid-free paper

Springer is part of Springer Science+Business Media (www.springer.com)

Preface

Classical Mechanics is a first step for those who begin to study Theoretical Physics. One can say that in this case, as in many other situations of life, the first step can be the most important one. In fact, the same subjects which are dealt with first in Mechanics are sometimes approached afterwards in more advanced courses such as Electrodynamics, Quantum Mechanics and Statistical Mechanics. Of course, the same concepts appear to be more sophisticated and sometimes more interesting when formulated in these frameworks. At the same time these formulations may also be more difficult. For this reason, the student who knows well Classical Mechanics can study other parts of Theoretical Physics easier and perhaps better.

The course of Classical Mechanics for Physics students is traditionally divided in two parts. The first one is technically simpler than the second one, which is also called Analytical Mechanics. In the second part, the mathematical and theoretical tools are more complicated and this allows us to consider the notions of Mechanics in a more general and profound way. At the same time, for many students, it is important to start the study of Analytical Mechanics with an appropriate preparation, and for this reason the one-semester introductory physics course is usually not enough. To address this situation, there is usually a course of Classical Mechanics, which is somehow intermediate between introductory Physics and Analytical Mechanics. Such an intermediate course is the subject of the present book. An important difference between such an introductory course and Analytical Mechanics is that the first one assumes a restricted use of mathematical methods and, therefore, leaves more room for physics intuition.

Classical Mechanics is known for centuries, but here we have introduced some relatively modern concepts, e.g., Einstein's equivalence principle. It turns out that they are useful also at the introductory level and, actually, help to solve some problems in a more simple way.

We believe that the textbooks directed to students should have the following important feature: they have to take into account the realities of academic life, when students have many disciplines to study in a short period of time. So the textbooks should be sufficiently short. In this book, we tried to be brief. The book is technical and aimed at clarifying the key notions and show how to solve problems of a standard level of difficulty, and sometimes a bit more difficult ones. Finally, we hope this book will be useful for the readers and wish everyone a good study of Classical Mechanics.

The first version of this book was published in Portuguese by "Livraria da Física" (USP, São Paulo) with a generous support from the research sector (PROPESQ) of Federal University of Juiz de Fora, the same is true for the first version of translation of the manuscript into English. We would like to acknowledge the efforts of Rafael Gonçalves, in typing the first version in Portuguese and also Tiago Mendes Rodrigues and Caique Affonso Garcia for contributing to the translation to English, respectively. We would like to thank all those colleagues and students who indicated to us misprints and shortcomings in the drafts and in the final Portuguese version, and especially Dra. Elena Konstantinova who made a very important contribution in improving and cleaning up the last version, which was translated to English. Finally, we acknowledge the contribution of Prof. Neil Ashby who professionally reviewed the book for Springer/NY and helped us to improve English and presentation.

Finally, we wish to thank CNPq, FAPEMIG and (I.Sh.) ICTP for the long-term support of our work.

Juiz de Fora, Brazil   Ilya L. Shapiro
                                                                              G. de Berredo-Peixoto

Contents

Notations

1. The notation for sum is

$$\sum_{i=1}^{N} x_i = x_1 + x_2 + \ldots + x_N.$$

2. The vectors are written in two different ways. In case of Latin letters, the "bold" style is used, e.g., **r**, **v** and **F**. For Greek letters we use arrows, e.g., $\vec{\omega}$ or $\vec{\alpha}$.

3. The scalar and vector product of two vectors, **a** and **b**, are written as

$$(\mathbf{a}, \mathbf{b}) = \mathbf{a} \cdot \mathbf{b} \quad \text{and} \quad [\mathbf{a}, \mathbf{b}] = \mathbf{a} \times \mathbf{b}$$

correspondingly. Both types of notations are common in the literature, so our goal is to give the students a chance to get used to these notations.

4. The modulus (magnitude or length) of a vector **a** is written as follows:

$$a = |\mathbf{a}| = \sqrt{\mathbf{a} \cdot \mathbf{a}} = \sqrt{a_1^2 + a_2^2 + a_3^2},$$

where a_1, a_2, a_3 are the components of the vector **a** in some orthonormal basis, for example, in Cartesian coordinates. The same notation is used for the radius-vector (or position vector) of a point particle,

$$\mathbf{r} = x\hat{\mathbf{i}} + y\hat{\mathbf{j}} + z\hat{\mathbf{k}}, \quad \text{that is,} \quad r = |\mathbf{r}| = \sqrt{x^2 + y^2 + z^2}.$$

5. The unit vector in the direction of **a** is denoted as **â**; this means that $\hat{\mathbf{a}} = \mathbf{a}/a$. Cartesian reference frame (system of coordinates) consists of the initial point (or origin) O, and the three orthonormal vectors $\hat{\mathbf{i}}, \hat{\mathbf{j}}, \hat{\mathbf{k}}$ which form the basis.

Chapter 1
Introduction

The Classical or Newtonian Mechanics deals with macroscopic movements. It is difficult to formulate all the foundations of Mechanics in a mathematically correct form; however, we can start from the following assumptions[1]:

1. Time and space are absolute. This means that clocks show the same amount of time anywhere regardless of their motion. It is important to clarify that here the clocks are seen as idealized instruments with absolute precision. The same applies to space measures. We assume, in this book, that the spatial measures of a given object do not change when in motion. A more general approach, where these postulates of Classical Mechanics are violated, is called Theory of Relativity.

2. The quantum effects are negligible. This can be seen, from the modern point of view, as an approximation, which is valid when we consider only length scales much larger than the typical size of an atom. For example, we can safely consider as macroscopic (i.e., objects studied by classical and not Quantum Mechanics) those structures which have linear sizes larger than 0.1 mm, but usually we can lower this limit by 1,000 or even 100,000 times. At the same time, even for sub-millimeter distances, some manifestations of quantum properties of vacuum can be observed. The corresponding phenomenon is called the Casimir Effect. For example, two parallel conducting planes, separated by a sub-millimeter distance in vacuum, experience an attractive force of quantum origin. This effect, however, does not influence, in most cases, the motion of particles or macroscopic bodies, which represent the main objects of our interest. In many cases atoms, nuclei and even electrons can be treated as classical objects, depending on the given physical problem. Of course, in order to decide whether Classical Mechanics applies to a given system or not, one should study the physical situation within the scope of Quantum Mechanics. In this book, we assume that Classical Mechanics is always appropriate. This means we assume that quantum effects simply do not take place.

3. Space is homogeneous and isotropic, time is homogeneous. The spatial homogeneity means that the properties of empty space are the same everywhere (there

---

[1] They are, from the standpoint of more general theories, approximations. We will discuss this later.

I.L. Shapiro and G. de Berredo-Peixoto, *Lecture Notes on Newtonian Mechanics*,
Undergraduate Lecture Notes in Physics, DOI 10.1007/978-1-4614-7825-6_1,
© Springer Science+Business Media, LLC 2013

is no privileged place in space). In the case of time, homogeneity means that the validity of the laws of physics does not change with time (from the standpoint of physical laws, there is no privileged moment of time). The isotropy of space means that it has the same properties in all directions. Of course, the properties of space can be changed by the presence of an external field. For example, in the presence of a homogeneous and uniform gravitational field the space is neither isotropic nor homogeneous. So when we talk about the properties of space, we assume the absence of external fields.

4. In three-dimensional space (hereafter called 3D) there are global basis vectors $\hat{\mathbf{i}}, \hat{\mathbf{j}}, \hat{\mathbf{k}}$. We chose these orthonormal vectors such that

$$\hat{\mathbf{i}} \cdot \hat{\mathbf{j}} = \hat{\mathbf{j}} \cdot \hat{\mathbf{k}} = \hat{\mathbf{k}} \cdot \hat{\mathbf{i}} = 0, \qquad \hat{\mathbf{i}} \cdot \hat{\mathbf{i}} = \hat{\mathbf{i}}^2 = \hat{\mathbf{j}}^2 = \hat{\mathbf{k}}^2 = 1.$$

More than that, we assume positively oriented (right handed) basis (see Fig. 1.1),

$$\hat{\mathbf{i}} \times \hat{\mathbf{j}} = \hat{\mathbf{k}}, \qquad \hat{\mathbf{k}} \times \hat{\mathbf{i}} = \hat{\mathbf{j}}, \qquad \hat{\mathbf{j}} \times \hat{\mathbf{k}} = \hat{\mathbf{i}}. \tag{1.1}$$

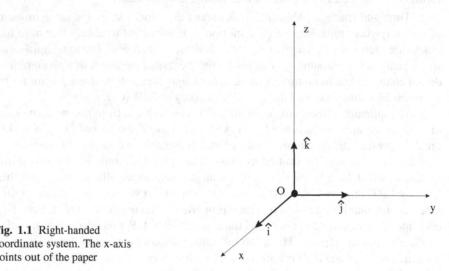

Fig. 1.1 Right-handed coordinate system. The x-axis points out of the paper

Any vector **b** can be represented as a linear combination of the basis vectors,

$$\mathbf{b} = \hat{\mathbf{i}} b_x + \hat{\mathbf{j}} b_y + \hat{\mathbf{k}} b_z.$$

This requirement means that the space is flat. In other words, the $3D$ space is the analogue of the $2D$ plane and not of a sphere or other curved surface where a global basis does not exist.

5. The speed of light and the speed of gravitational signals are infinite. This means that when a planet (e.g., the Moon) changes its position, another planet (e.g., the Earth) will "feel" the gravitational force of the first planet in its new position instantly. In other words, the acceleration vector of the Earth is always directed

to the instantaneous position of the Moon and not to its preceding position. This hypothesis is necessary in Classical Mechanics, but it is actually only valid for bodies whose speed of motion is much less than the speed of light, $c \approx 300,000$ km/s.

We can discuss a little more the validity of assumptions 1–5. As it was already mentioned, in fact all of them are not absolutely correct. The statements 1 and 5 are violated in the theory of special relativity, the assertion 2 is "violated" in the framework of Quantum Mechanics, the statements 3 and 4 are "violated" by General Relativity. In other words, for some physical situations, the space is not flat, time and space are not absolute, the notions of homogeneity and isotropy must be applied carefully and, of course, the speed of light is not infinite. The natural question is: "Does this mean that Classical Mechanics is a wrong science?". The answer is no. To understand how this is possible, let us imagine the following situation: Somebody comes to Copacabana beach and pours a bucket of water in the sea. Will the sea level, from the standpoint of physics, increase? Of course not. But why? Has the amount of water in the sea increased? Yes, it has. However, in physics we say that something exists only when it can be measured. Such a measurement may be a very distant possibility, very far from experimental reality, but this possibility must exist, in principle. In the case of a bucket of water, the sea level can only be defined with some precision, taking into account waves, swimmers, fish and, maybe, a capybara that fell overboard by accident. There is not, even theoretically, a measure of the increase in the sea level in Copacabana because of an extra bucket of water. Therefore, the sea level does not increase, from the physics viewpoint. Almost the same applies to the position of the Moon that was discussed earlier. With great precision (considered perfect in the field of Classical Mechanics), the movement of the Moon is so slow that it does not influence the gravitational interaction. For the description of many mechanical phenomena we do not need to take into account the finite speed of light or quantum effects. Ultimately, in the framework of given approximations, Classical Mechanics is a very successful science. Moreover, as already mentioned above, it serves as a basis for more advanced sections of theoretical physics.

The main objects of study of Classical Mechanics are as follows:
- A material point (also called particle or point mass), which is a body with negligible size;
- A rigid body. This means a body with all the distances between its constituent points completely fixed.

An important observation is that the same body can be legitimately regarded as a point mass in one situation, as a rigid body in a different situation and as an object that cannot be treated in either of these two ways, in a third situation. One possible example is the Earth. When we study its motion around the Sun, it can be seen as a point. When we study Earth rotation around its axis, it can be considered as a rigid body. But when we are interested in the study of tides because of its interaction with Moon and Sun, both approaches are no longer suitable and one has to use another framework to describe these phenomena. We leave to the reader as an exercise to find other similar examples.

Throughout this course, consultation with other books of Classical Mechanics can be useful for the student's learning. It is not possible to cite all Mechanics books that address the subject at a similar level to the approach presented here, but we can mention some of them, Refs. [5, 8, 14, 15, 18, 19, 21, 24, 26, 27, 31, 32]. Most of these books include also Analytical Mechanics. The student interested in Analytical Mechanics can find many other books, among which we cite only some of the best known in Refs. [10, 11, 20, 30], besides books with a more mathematical approach, such as Refs. [1, 3, 9].

Chapter 2
Kinematics

Abstract Kinematics is focused on the mathematical description of motion and does not concern its dynamical causes. We shall consider the motion of particles in different coordinate systems, including Cartesian, polar (in two-dimensional $2D$ case), cylindrical, spherical and also in the co-moving basis which is useful in reducing $3D$ motion to $2D$. Also we are going to deal with the kinematics of rigid bodies, which can be decomposed into a translational motion and a rotation around an instantaneous axis. Finally, we will consider the motion in an accelerated reference frame and introduce the notion of covariant derivative which is helpful, in particular, for the kinematics in rotating coordinates.

2.1 Kinematics of a Particle

To parameterize the motion of a material point, we use clocks and a space coordinate system. We assume that all clocks (idealized devices that measure time), regardless of their positions, velocities or accelerations, indicate the same measure of time. The coordinate system (also called reference system or reference frame) is used to parameterize the three-dimensional space (for the sake of brevity, we use the notation $3D$, or $2D$ for the two-dimensional plane, or $1D$ for one-dimensional space). As a first step, we define the Cartesian coordinates. This coordinate system includes the point of origin O, and the orthonormal basis $(\hat{\mathbf{i}},\hat{\mathbf{j}},\hat{\mathbf{k}})$. These three vectors must have right-handed orientation (see Fig. 1.1), i.e., they must satisfy the relations (1.1).

For any point M we can define the position vector (also called radius-vector) $\mathbf{OM} = \mathbf{r} = \hat{\mathbf{i}}x + \hat{\mathbf{j}}y + \hat{\mathbf{k}}z$. The coefficients (x, y, z) are called coordinates of the point in space. In the case of a laboratory reference frame, we assume that the vectors $\hat{\mathbf{i}}$, $\hat{\mathbf{j}}$ and $\hat{\mathbf{k}}$ are constant and static (Fig. 2.1).

For a material point (also called a particle or point-like body) in motion, the coordinates depend on time:

$$\mathbf{r}(t) = x(t)\hat{\mathbf{i}} + y(t)\hat{\mathbf{j}} + z(t)\hat{\mathbf{k}}. \qquad (2.1)$$

I.L. Shapiro and G. de Berredo-Peixoto, *Lecture Notes on Newtonian Mechanics*, Undergraduate Lecture Notes in Physics, DOI 10.1007/978-1-4614-7825-6_2, © Springer Science+Business Media, LLC 2013

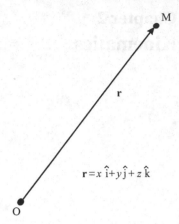

Fig. 2.1 Position vector (or radius-vector) in the Cartesian basis

$\mathbf{r}=x\,\hat{\mathbf{i}}+y\hat{\mathbf{j}}+z\,\hat{\mathbf{k}}$

The relation above describes the particle motion, i.e., it shows how the position vector depends on time.

In some cases we are not interested in this dependence and just want to know the curve of spatial motion of the particle, called a trajectory. In fact, the relationship (2.1) is a particular case of the trajectory equation, where time t serves as a parameter. When the time dependence does not matter, we can choose another parameter. In this case, there are two special options, namely:

1) We can use one of the coordinates as a parameter. For example, consider the following motion:

$$x = vt\cos\alpha, \qquad y = 0, \qquad z = vt\sin\alpha - \frac{gt^2}{2}, \qquad (2.2)$$

where v, α, g are constants. To obtain the trajectory parameterized by x, we need to solve the first equation in (2.2) with respect to t and replace this solution into the third equation. Thus, we obtain $t = x/(v\cos\alpha)$ and the equation of the trajectory is cast into the form

$$y = 0, \qquad z = x\tan\alpha - \frac{gx^2}{2v^2\cos^2\alpha}. \qquad (2.3)$$

The system of Eqs. (2.3) has one less equation compared to (2.2). This reflects the fact that some information has been lost on the way, namely the Eqs. (2.3) do not contain information about the time dependence. They only show the spatial line, in other words they contain geometrical information about the movement, but not the dynamical content.

2) We can use the natural parameter, l, along the trajectory. If we choose some point on the curve (trajectory) and define it as a starting point, the natural parameter represents the distance along the curve between the starting point and the point of our interest. To set this parameter, we consider an infinitesimal displacement along the curve. We have

$$dl^2 = dx^2 + dy^2 + dz^2,$$

where (x, y, z) satisfy the equation of the trajectory. Correspondingly, if the starting point is A and the point of interest is M, the value of the parameter l is given by the line integral of first kind (see Mathematical Appendix for more details)

$$l = \int\limits_{(AM)} dl. \tag{2.4}$$

We assume that initially the path is parameterized by some parameter λ such that $\mathbf{r} = \mathbf{r}(\lambda)$. In this case, the integral (2.4) can be written as a Riemann integral (definite integral)

$$l = \int\limits_{\lambda_A}^{\lambda_M} \sqrt{\left(\frac{dx}{d\lambda}\right)^2 + \left(\frac{dy}{d\lambda}\right)^2 + \left(\frac{dz}{d\lambda}\right)^2}\, d\lambda,$$

where λ_A and λ_M are the values that correspond to the points A and M, correspondingly. For example, when $\lambda = x$, the last equation is written as

$$l = \int\limits_{x_A}^{x_M} \sqrt{1 + \left(\frac{dy}{dx}\right)^2 + \left(\frac{dz}{dx}\right)^2}\, dx.$$

Another interesting case, when $\lambda = t$ (time) will be discussed later on.

A major advantage in using the natural parameter along the curve (also called the canonical parameter) is that a change in the position of the particle in the trajectory always corresponds to a change of the natural parameter. None of the other parameters considered above (e.g., coordinate x or time t) have this property. At the same time, our main objective is the study of the dynamics of particles, so the dependence on time plays a key role in our considerations.

The velocity (instantaneous velocity) of the particle is defined as the time derivative of the position vector,

$$\mathbf{v}(t) = \lim_{\Delta t \to 0} \frac{\Delta \mathbf{r}}{\Delta t} = \lim_{\Delta t \to 0} \frac{\mathbf{r}(t + \Delta t) - \mathbf{r}(t)}{\Delta t},$$

or

$$\mathbf{v}(t) = \dot{\mathbf{r}}(t) = \frac{d\mathbf{r}}{dt} = \hat{\mathbf{i}}\frac{dx}{dt} + \hat{\mathbf{j}}\frac{dy}{dt} + \hat{\mathbf{k}}\frac{dz}{dt} = \dot{x}\hat{\mathbf{i}} + \dot{y}\hat{\mathbf{j}} + \dot{z}\hat{\mathbf{k}}. \tag{2.5}$$

Henceforth, we will adopt the dot over the quantity to denote its time derivative. To determine the magnitude of velocity, we can use the theorem of Pythagoras, as

$$v = |\mathbf{v}| = \sqrt{\dot{x}^2 + \dot{y}^2 + \dot{z}^2}.$$

It is easy to see that this is nothing else but the time derivative of the natural parameter along the trajectory,

$$v = \frac{\sqrt{dx^2 + dy^2 + dz^2}}{dt} = \frac{dl}{dt}. \tag{2.6}$$

Dividing the velocity vector by its magnitude, we find the unit vector

$$\hat{\mathbf{v}} = \frac{\mathbf{v}}{v}$$

which is directed parallel to the tangent line of the trajectory in a given point and points to the side of growth of the parameter l.

We can invert the relation (2.5). Integrating $\mathbf{v}(t)$ and taking into account the initial condition $\mathbf{r}(t_0) = \mathbf{r}_0$, we obtain

$$\mathbf{r}(t) = \mathbf{r}_0 + \int_{t_0}^{t} \mathbf{v}(\tau) d\tau.$$

For example, if $\mathbf{v}(t) = \mathbf{v}_0 = const$, we have

$$\mathbf{r}(t) = \mathbf{r}_0 + \mathbf{v}_0(t - t_0).$$

In order to simplify the formulas, we can assume that $t_0 = 0$. In this case, for example, $\mathbf{r}(t) = \mathbf{r}_0 + \mathbf{v}_0 \cdot t$.

Now we define the instantaneous acceleration, $\mathbf{a}$, as the time derivative of the velocity vector,

$$\mathbf{a} = \dot{\mathbf{v}} = \frac{d\mathbf{v}}{dt} = \dot{v}_x \hat{\mathbf{i}} + \dot{v}_y \hat{\mathbf{j}} + \dot{v}_z \hat{\mathbf{k}}, \tag{2.7}$$

or, with respect to the position vector,

$$\mathbf{a} = \ddot{\mathbf{r}} = \frac{d^2 \mathbf{r}}{dt^2} = \ddot{x} \hat{\mathbf{i}} + \ddot{y} \hat{\mathbf{j}} + \ddot{z} \hat{\mathbf{k}}.$$

Obviously,

$$a_x = \dot{v}_x = \ddot{x}, \qquad a_y = \dot{v}_y = \ddot{y}, \qquad a_z = \dot{v}_z = \ddot{z}.$$

In the general case, the components of acceleration depend on time. We can integrate the relation (2.7) using the initial condition $\mathbf{v}(t = 0) = \mathbf{v}_0$,

$$\mathbf{v}(t) = \mathbf{v}_0 + \int_0^t \mathbf{a}(\tau) d\tau.$$

In particular, for $\mathbf{a} \equiv \mathbf{a}_0 = const$, we have

$$\mathbf{v}(t) = \mathbf{v}_0 + \mathbf{a}_0 \int_0^t d\tau = \mathbf{v}_0 + \mathbf{a}_0 t. \tag{2.8}$$

In this case, for the radius-vector we find

$$\mathbf{r} = \mathbf{r}_0 + \mathbf{v}_0 t + \frac{\mathbf{a}_0 t^2}{2}.$$

In the general case of an arbitrary time-dependent acceleration, one has a more general expression,

$$\mathbf{r}(t) = \mathbf{r}_0 + \int_0^t d\tau\, \mathbf{v}(\tau) = \mathbf{r}_0 + \int_0^t d\tau \left\{ \mathbf{v}_0 + \int_0^\tau \mathbf{a}(\tau')d\tau' \right\}$$

$$= \mathbf{r}_0 + \mathbf{v}_0 t + \int_0^t d\tau \int_0^\tau d\tau'\, \mathbf{a}(\tau'). \tag{2.9}$$

As an example, consider the motion of a point-like body in a homogeneous and uniform gravitational field. In this case, the acceleration is constant, $\mathbf{a} \equiv \mathbf{g} = -g\hat{\mathbf{k}}$. We can always choose the reference frame such that the vector basis satisfies $\hat{\mathbf{j}} \perp \mathbf{v}_0$. In this case, we can write

$$\mathbf{v}_0 = v_{ox}\hat{\mathbf{i}} + v_{oz}\hat{\mathbf{k}} = v_0 \left(\hat{\mathbf{i}}\cos\alpha + \hat{\mathbf{k}}\sin\alpha \right),$$

where α is the angle between $\mathbf{v}_0$ and $\hat{\mathbf{i}}$. Using the previous results (2.8) and (2.9), we can write the solution

$$\mathbf{v}(t) = v_{ox}\hat{\mathbf{i}} + \left(v_{oz} - gt \right)\hat{\mathbf{k}},$$

$$\mathbf{r} = \mathbf{r}_0 + v_{ox}t\hat{\mathbf{i}} + \left(v_{oz}t - \frac{gt^2}{2} \right)\hat{\mathbf{k}},$$

i.e., in components,

$$x = x_0 + v_{ox}t = x_0 + v_0 t\cos\alpha, \qquad y = y_0,$$

$$z = z_0 + v_0 t\sin\alpha - \frac{gt^2}{2}.$$

The last formula describes the motion of a particle in a homogeneous and uniform gravitational field. Eliminating the time t, we obtain the equation of the trajectory (2.3) for the case $x_0 = y_0 = z_0 = 0$.

As far as the description of the motion of a material point involves the radius-vector, velocity or acceleration, a relation between them represents, typically, a differential equation. The solution of this equation provides the kinematical description by a vector function $\mathbf{r}(t)$.

Let us consider a simple example of this sort. Suppose we know that for some particle $\mathbf{a} = -k\mathbf{v}$, with $k = const$ and $k > 0$. The dynamics of the point is completely defined by this equation and by the necessary initial data (in mathematical terminology, this combination is called the Cauchy problem),

$$\frac{d\mathbf{v}}{dt} = -k\mathbf{v} \qquad \text{and} \qquad \mathbf{v}(t=0) = \mathbf{v}_0. \tag{2.10}$$

The first formula represents a first order differential equation with respect to the velocity **v**, and the second one is an initial condition. In fact, it is a system of three equations for v_x, v_y, v_z. One can solve this problem, e.g., in arbitrary Cartesian coordinates. At the same time, the problem can be essentially simplified if we choose a special basis. Let us introduce the orthonormal basis consisting of the three vectors $\hat{\mathbf{v}}, \hat{\mathbf{n}}_1$ and $\hat{\mathbf{n}}_2$, where $\hat{\mathbf{v}} = \mathbf{v}/v$, also $\hat{\mathbf{n}}_{1,2} \perp \hat{\mathbf{v}}$ and $\hat{\mathbf{n}}_1 \perp \hat{\mathbf{n}}_2$. It is easy to see that the direction of velocity $\hat{\mathbf{v}}$ remains constant, and so we can write $\mathbf{v}(t) = \hat{\mathbf{v}} \cdot v(t)$. The first equation (2.10) becomes

$$\frac{d\mathbf{v}}{dt} = \hat{\mathbf{v}}\frac{dv}{dt} = -k\hat{\mathbf{v}}v.$$

Finally, we have the only one differential equation, $\dot{v}(t) = -kv(t)$. To solve it, we will use the method of separation of variables. Grouping all terms involving v on the left side and the terms involving t on the right side, we write

$$\frac{dv}{v} = -kdt.$$

Assuming that $v(t)$ is a function that satisfies the last relation and integrating both sides, we arrive at the formula

$$\ln\left|\frac{v}{C}\right| = -kt, \qquad (2.11)$$

where C is a constant of integration. To define this constant, one has to use the initial condition. Substituting $t = 0$ and $v = v_0$ in Eq. (2.11), we find $\ln|v_0/C| = 0$, i.e., $C = v_0$. Finally, the solution for the velocity is

$$\mathbf{v} = \hat{\mathbf{v}}v_0 e^{-kt} = \mathbf{v}_0 e^{-kt}.$$

It is easy to see that the last formula satisfies both relations (2.10). In the limit $t \to \infty$ we have $\mathbf{v} \to 0$. To find the kinematical equations, it is necessary to introduce a new initial condition $\mathbf{r}(0) = \mathbf{r}_0$, and use the general prescription (2.9),

$$\mathbf{r} = \mathbf{r}_0 + \hat{\mathbf{v}}v_0 \int_0^t e^{-k\tau}d\tau$$

$$= \mathbf{r}_0 + \mathbf{v}_0 \cdot \left(-\frac{1}{k}\right)e^{-k\tau}\Big|_0^t = \mathbf{r}_0 + \frac{\mathbf{v}_0}{k}(1 - e^{-kt}).$$

In the limit $t \to \infty$, we find

$$\mathbf{r} \to \mathbf{r}_0 + \frac{\mathbf{v}_0}{k}.$$

The last example shows the great importance of choosing a useful reference frame. Here we uses a basis related to the motion of the particle itself and got a real simplification in the solution. We will develop the same idea in a more general form in the next section.

Exercise 1. A particle moves in space with coordinates x, y, z, according to the equations of motion

$$x = 3vt, \quad y = 3L\cos\omega t, \quad z = -5L\sin\omega t.$$

Write the same equations in vector form and find the velocity $\mathbf{v}$ and acceleration $\mathbf{a}$ as functions of time.

Answers:
$$\mathbf{r} = 3vt\hat{\mathbf{i}} + 3\hat{\mathbf{j}}L\cos\omega t - 5\hat{\mathbf{k}}L\sin\omega t;$$
$$\mathbf{v} = 3v\hat{\mathbf{i}} - 3\omega\hat{\mathbf{j}}L\sin\omega t - 5\omega\hat{\mathbf{k}}L\cos\omega t;$$
$$\mathbf{a} = -3\omega^2\hat{\mathbf{j}}L\cos\omega t + 5\omega^2\hat{\mathbf{k}}L\sin\omega t.$$

2.2 Motion of a Particle in the Co-moving Basis

Let us introduce the so-called co-moving basis, i.e., the one which is attached to the particle. The advantage of this basis compared to some static basis, $\hat{\mathbf{i}}, \hat{\mathbf{j}}, \hat{\mathbf{k}}$, is related to the fact that, in many situations, only two vectors of the co-moving basis are relevant. In many cases the laws of mechanics involve the velocity $\mathbf{v}$ and the acceleration $\mathbf{a}$ vectors of the particle. Then we can always choose a single plane in which the motion occurs.[1] This plane should include both vectors, $\mathbf{v}$ and $\mathbf{a}$. Now we can build the corresponding orthonormal basis. The first vector of this basis is $\hat{\mathbf{v}} = \mathbf{v}/v$, this means it has the same direction as velocity. To choose the second vector, we assume that the vectors $\mathbf{v}$ and $\mathbf{a}$ are not parallel. In this case one can choose the unit vector $\hat{\mathbf{n}}$, which is normal to the trajectory, such that
  (i) $\hat{\mathbf{v}} \perp \hat{\mathbf{n}}$;
 (ii) $\hat{\mathbf{n}}$ belongs to the same plane as the vectors $\mathbf{v}$ and $\mathbf{a}$;
(iii) The plane is divided into two semi-planes by the line that contains the vector $\mathbf{v}$. The vector $\hat{\mathbf{n}}$ lies in the same half-plane that the vector $\mathbf{a}$.
Figure 2.2 shows the positions of the vectors defined above.
  It is important to remember that the vectors $\hat{\mathbf{v}}$ and $\hat{\mathbf{n}}$, in general, depend on time, such that $\hat{\mathbf{v}} = \hat{\mathbf{v}}(t)$ and $\hat{\mathbf{n}} = \hat{\mathbf{n}}(t)$. It is clear that this dependence does not concern their absolute values, but only the directions of the two vectors.
  Let us consider the velocity and acceleration vectors on the co-moving basis. The velocity can be written as

$$\mathbf{v}(t) = v(t) \cdot \hat{\mathbf{v}}(t). \tag{2.12}$$

For the sake of convenience, in what follows we shall omit the time argument. For acceleration, one can write

---

[1] The particular case where $\mathbf{v}\|\mathbf{a}$ does not require the basis that we are discussing now and can be treated in the same manner as we did in the example discussed above, at the end of the previous section.

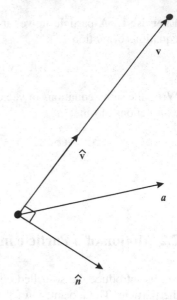

Fig. 2.2 Construction of the co-moving basis. All the vectors belong to the same plane

$$\mathbf{a} = a_t \cdot \hat{\mathbf{v}} + a_n \cdot \hat{\mathbf{n}}. \tag{2.13}$$

The components of acceleration on the co-moving base have the following nomenclature: a_t is the tangential acceleration and a_n is the normal acceleration. We can establish some useful relations between the terms that appear in the formula (2.13). First, $a^2 = a_t^2 + a_n^2$.

Taking the time derivative of Eq. (2.12), we obtain

$$\mathbf{a} = \frac{dv}{dt} \cdot \hat{\mathbf{v}} + v \cdot \frac{d\hat{\mathbf{v}}}{dt}. \tag{2.14}$$

Comparing (2.13) and (2.14), we can see that the tangential component of acceleration, a_t, is the time derivative of the absolute value of the velocity vector,

$$a_t = \frac{dv}{dt} = \frac{d^2 l}{dt^2}. \tag{2.15}$$

We know that (see Fig. 2.3)

$$\frac{d\hat{\mathbf{v}}}{dt} = \lim_{\Delta t \to 0} \frac{\Delta \hat{\mathbf{v}}}{\Delta t} = \lim_{\Delta t \to 0} \frac{\hat{\mathbf{v}}(t + \Delta t) - \hat{\mathbf{v}}(t)}{\Delta t}$$

and, at the same time, the magnitude of the vector $\hat{\mathbf{v}}$ remains constant, $|\hat{\mathbf{v}}| \equiv 1$. Obviously, this is only possible in the case

$$\frac{d\hat{\mathbf{v}}}{dt} \perp \hat{\mathbf{v}}, \qquad \text{that is,} \qquad \frac{d\hat{\mathbf{v}}}{dt} \parallel \hat{\mathbf{n}}.$$

Finally we can write Eq. (2.14) in the following way:

$$v\frac{d\hat{\mathbf{v}}}{dt} = \mathbf{a} - \dot{v}\hat{\mathbf{v}}, \qquad \text{that is,} \qquad a_n\hat{\mathbf{n}} = v\frac{d\hat{\mathbf{v}}}{dt}.$$

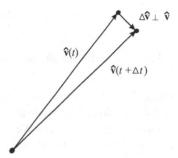

Fig. 2.3 Vector representation
of the vectors $\Delta\mathbf{v}$ and $\mathbf{v}$

At this stage we can use the natural parameter, l, to define one more important characteristic of the trajectory, called the radius of curvature, R. As a next step, we will find a relationship between this quantity and the normal component of acceleration. The value of R is defined by the relation

$$\frac{d\hat{\mathbf{v}}}{dl} = \frac{\hat{\mathbf{n}}}{R}. \tag{2.16}$$

It is clear that R does not necessarily have the same value at different points of the trajectory. At the same time, the radius of curvature is a geometric feature, i.e., it depends only on the trajectory and not on the given particle's motion along it. This means that it does not matter if the particle passes through a given point of the trajectory fast or slow, R is the same. At the same time, one can establish an interesting relation between the normal component of acceleration, the magnitude of the velocity and the radius of curvature,

$$a_n = \frac{v^2}{R}. \tag{2.17}$$

To demonstrate this relation, we write again

$$\mathbf{v} = \hat{\mathbf{v}}\cdot v, \qquad v = \frac{dl}{dt}, \qquad a_t = \frac{dv}{dt}$$

and

$$\mathbf{a} = \frac{d\mathbf{v}}{dt} = \hat{\mathbf{v}}\cdot\frac{dv}{dt} + a_n\cdot\hat{\mathbf{n}}. \tag{2.18}$$

Let us consider, on the other hand,

$$\frac{d\mathbf{v}}{dt} = \frac{d}{dt}(\hat{\mathbf{v}}\cdot v) = \frac{d\hat{\mathbf{v}}}{dt}v + \hat{\mathbf{v}}\frac{dv}{dt} = \frac{d\hat{\mathbf{v}}}{dt}v + \hat{\mathbf{v}}\cdot a_t. \tag{2.19}$$

Comparing (2.18) with (2.19), we have

$$\hat{\mathbf{v}}a_t + v\frac{d\hat{\mathbf{v}}}{dt} = \hat{\mathbf{v}}a_t + a_n \cdot \hat{\mathbf{n}}, \qquad \text{so that} \qquad \frac{d\hat{\mathbf{v}}}{dt} = \frac{a_n\,\hat{\mathbf{n}}}{v}. \tag{2.20}$$

Finally, we have

$$\frac{d\hat{\mathbf{v}}}{dt} = \frac{d\hat{\mathbf{v}}}{dl} \cdot \frac{dl}{dt} = v\frac{d\hat{\mathbf{v}}}{dl}. \tag{2.21}$$

Taking into account the definition (2.16), the Eqs. (2.21) and (2.20) give the Eq. (2.17).

Example 1. To illustrate the physical meaning of R, consider a particular case of circular motion of a particle along a circumference of radius R with constant angular velocity ω (See Fig. 2.4). We choose the initial instant $t = 0$ in which the particle has coordinates $x(0) = R$ and $y(0) = 0$.

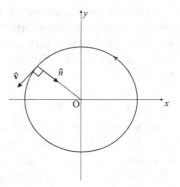

Fig. 2.4 Uniform circular motion

In this case, we have

$$x = R\cos\omega t, \qquad y = R\sin\omega t,$$

or, in vector form,

$$\mathbf{r} = R\left(\hat{\mathbf{i}}\cos\omega t + \hat{\mathbf{j}}\sin\omega t\right).$$

For the velocity vector, we obtain

$$\mathbf{v} = \frac{d\mathbf{r}}{dt} = R\omega\left(-\hat{\mathbf{i}}\sin\omega t + \hat{\mathbf{j}}\cos\omega t\right),$$

It is easy to see that $r = |\mathbf{r}| = R = const$ and also $v = |\mathbf{v}| = \omega R = const$. For acceleration, we have

$$\mathbf{a} = \frac{d\mathbf{v}}{dt} = -R\omega^2\left(\hat{\mathbf{i}}\cos\omega t + \hat{\mathbf{j}}\sin\omega t\right) = \hat{\mathbf{n}}R\omega^2 = \frac{v^2\,\hat{\mathbf{n}}}{R}.$$

One can see that in this case $a_t = 0$ and $a_n = v^2/R$. The acceleration vector is directed to the center. The curvature radius is exactly the circumference radius. In the case of circular motion, the normal component of acceleration is called centripetal acceleration. The normal component is not always the only component of acceleration. The tangential component of the acceleration is related to the angular acceleration, β, as follows:

$$a_t = \beta R, \quad \text{where} \quad \beta = \frac{d\omega}{dt}. \tag{2.22}$$

In the general case, when the movement is not necessarily circular, for each point of the trajectory we can imagine a circumference with radius R such that a_n is the centripetal acceleration corresponding to this circle (see illustration in Fig. 2.5). For any motion, the magnitude of the acceleration is

$$|\mathbf{a}| = \sqrt{a_t^2 + a_n^2} = \sqrt{\left(\frac{dv}{dt}\right)^2 + \left(\frac{v^2}{R}\right)^2}$$

and the direction of $\mathbf{a}$ depends on the relation between a_t and a_n. For example, if $a_t = 0$, the acceleration has only a normal component and the absolute value of the velocity is constant.

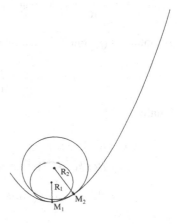

Fig. 2.5 R_1 and R_2 are the radii of curvature at the local respective points M_1 and M_2 of the trajectory

Example 2. A particle is in circular motion such that its angular acceleration is positive and proportional to its angular velocity. Its initial angular velocity is ω_0. Find the direction of acceleration as a function of time.

Solution. We consider the angle of rotation as a function of time. We have

$$\omega(t) = \frac{d\varphi(t)}{dt}, \quad \beta(t) = \frac{d\omega(t)}{dt} = k\omega(t),$$

where k is some coefficient. We choose the initial condition $\varphi(0) = \varphi_0$. As the first step, we solve the differential equation for $\omega(t)$, using the method of separation of variables. Thus,

$$\frac{d\omega}{\omega} = kdt \quad \Rightarrow \quad \ln\left|\frac{\omega}{\omega_0}\right| = kt,$$

where we already used the initial condition $\omega(0) = \omega_0$. So, $\omega(t) = \omega_0 e^{kt}$. This means that

$$v(t) = R\omega(t) = R\omega_0 e^{kt}$$

and also

$$a_t = \frac{dv}{dt} = R\omega_0 k e^{kt} = R\beta(t).$$

For the normal component of acceleration, we find

$$a_n = \frac{v^2}{R} = R\omega_0^2 e^{2kt}.$$

To find the direction of the acceleration, we remember that

$$\hat{\mathbf{n}} = -\frac{\mathbf{r}}{R} = -\left(\hat{\mathbf{i}}\cos\varphi + \hat{\mathbf{j}}\sin\varphi\right) \quad \text{and} \quad \hat{\mathbf{v}} = -\hat{\mathbf{i}}\sin\varphi + \hat{\mathbf{j}}\cos\varphi.$$

To obtain $\varphi(t)$, one needs to integrate $\omega(t)$:

$$\varphi(t) = \varphi_0 + \int_0^t \omega(t)dt = \varphi_0 + \frac{\omega_0}{k}\left(e^{kt} - 1\right). \tag{2.23}$$

Finally,

$$\begin{aligned}
\mathbf{a} &= a_t \cdot \hat{\mathbf{v}} + a_n\hat{\mathbf{n}} \\
&= R\omega_0 k e^{kt} \cdot \left(-\hat{\mathbf{i}}\sin\varphi + \hat{\mathbf{j}}\cos\varphi\right) - R\omega_0^2 e^{2kt} \cdot \left(\hat{\mathbf{i}}\cos\varphi + \hat{\mathbf{j}}\sin\varphi\right) \\
&= R\omega_0 k e^{kt}\left[-\hat{\mathbf{i}}\left(\sin\varphi + \frac{\omega_0 e^{kt}}{k}\cos\varphi\right) + \hat{\mathbf{j}}\left(\cos\varphi - \frac{\omega_0 e^{kt}}{k}\sin\varphi\right)\right],
\end{aligned}$$

where φ is given by the relation (2.23). This formula defines the magnitude and direction of acceleration $\mathbf{a}$. To find the unit direction of $\mathbf{a}$, we normalize it,

$$\hat{\mathbf{a}} = \frac{\mathbf{a}}{a} = \frac{-\hat{\mathbf{i}}\left(\sin\varphi + \omega_0 k e^{kt}\cos\varphi\right) + \hat{\mathbf{j}}\left(\cos\varphi - \omega_0 k e^{kt}\sin\varphi\right)}{\sqrt{1 + \frac{\omega_0^2 e^{2kt}}{k^2}}}.$$

Exercises

1. A projectile is launched from the origin O with the initial velocity which has magnitude v_0 and makes angle α with the horizon. Consider a straight line which passes through the origin and makes an angle β with the horizon, such that $\beta < \alpha$. After some time, the projectile crosses the line at the point P. What is the choice of α such that the distance between P and the origin be a maximal one?

Solution. The equation for the trajectory and the equation of the straight line are respectively

$$y = x\tan\alpha - \frac{gx^2}{2v_0^2\cos^2\alpha} \qquad \text{and} \qquad y = x\tan\beta. \qquad (2.24)$$

By replacing the second equation in (2.24) into the first one, we obtain the coordinate x of the point P:

$$x = (\tan\alpha - \tan\beta)\frac{2v_0^2\cos^2\alpha}{g}.$$

As far as x corresponds to the distance between P and the origin, multiplied by $\cos\beta$, the maximum value of x will result in a maximum for that distance. So, one only needs to find out what is the value of α that maximizes x. Taking the derivative $dx/d\alpha = 0$ (or, simpler, with respect to $t = \tan\alpha$), after a small algebra we arrive at the solution for the angle,

$$\alpha = \arctan\left(\frac{\sin\beta + 1}{\cos\beta}\right) = \frac{\pi}{4} + \frac{\beta}{2}.$$

2. A small rubber ball is dropped from rest at a height h over the ground and falls vertically. It bounces from the ground and returns to the height $h/2$. Every time ball hits the ground, it returns to half the previous height.

(a) Under these conditions, will it be possible to observe a finite duration of time τ, in which the ball stops on the floor? If so, calculate τ.

(b) Considering the interval of time between the instant in which the ball is released and the instant when it stops on the floor, calculate the magnitude of its average velocity and the average of absolute value of velocity (note that there is a difference between these two quantities).

Answers: $\tau = \sqrt{\frac{2h}{g}}(2\sqrt{2}+3)$, $\quad |\langle\mathbf{v}\rangle| = \frac{h}{\tau}$, $\quad \langle v\rangle = \frac{3\sqrt{gh}}{2}(3\sqrt{2}-4)$.

3. For the motion of particle on the $2D$ plane in the co-moving reference frame, prove the relation which is dual to (2.16),

$$\frac{d\hat{n}}{dl} = -\frac{\hat{v}}{R}. \qquad (2.25)$$

Discuss whether this formula holds for the general $3D$ motion.

2.3 Polar Coordinates in the Plane, Cylindrical and Spherical in the Space

Depending on the problem, the use of Cartesian coordinates is not always the most useful. Sometimes it is better to work in some other reference frame, for example, in the co-moving coordinates discussed above. Almost always we want to use coordinates that correspond to an orthogonal and normalized (orthonormal) basis. In this section we review some standard and most frequently used choices. As a first example, let us consider polar coordinates on the plane.

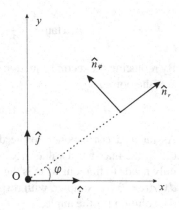

Fig. 2.6 Cartesian and polar bases

The polar coordinates (see Fig. 2.6) r and φ can be defined by the relations

$$\begin{cases} x = r\cos\varphi \\ y = r\sin\varphi. \end{cases}$$

To use a vector formalism, we define unit basis vectors

$$\hat{\mathbf{n}}_r = \frac{\mathbf{r}}{r} = \hat{\mathbf{i}}\cos\varphi + \hat{\mathbf{j}}\sin\varphi,$$
$$\hat{\mathbf{n}}_\varphi = -\hat{\mathbf{i}}\sin\varphi + \hat{\mathbf{j}}\cos\varphi. \qquad (2.26)$$

Exercise 2. Verify that the vectors (2.26) are orthonormal, that means the relations

$$\hat{\mathbf{n}}_r \cdot \hat{\mathbf{n}}_r = \hat{\mathbf{n}}_\varphi \cdot \hat{\mathbf{n}}_\varphi = 1 \quad \text{and} \quad \hat{\mathbf{n}}_r \cdot \hat{\mathbf{n}}_\varphi = 0,$$

where the dot indicates the scalar product.

Now we can imagine a particle with polar coordinates r, φ such that $r = r(t)$ and $\varphi = \varphi(t)$. Our next goal is to calculate the velocity and acceleration of the material point in polar coordinates. As the first step, we compute the time derivatives of the basis vectors (2.26),

$$\frac{d\hat{\mathbf{n}}_r}{dt} \quad \text{and} \quad \frac{d\hat{\mathbf{n}}_\varphi}{dt}.$$

It is easy to obtain, using Eqs. (2.26),

$$\frac{d\hat{\mathbf{n}}_r}{dt} = \dot{\varphi}(-\hat{\mathbf{i}}\sin\varphi + \hat{\mathbf{j}}\cos\varphi) = \dot{\varphi}\hat{\mathbf{n}}_\varphi,$$

$$\frac{d\hat{\mathbf{n}}_\varphi}{dt} = \dot{\varphi}(-\hat{\mathbf{i}}\cos\varphi - \hat{\mathbf{j}}\sin\varphi) = -\dot{\varphi}\hat{\mathbf{n}}_r. \tag{2.27}$$

Now we are at the point of writing the velocity and acceleration in polar coordinates. From Eqs. (2.27) we obtain

$$\mathbf{v} = \frac{d\mathbf{r}}{dt} = \frac{d}{dt}(r\hat{\mathbf{n}}_r) = \dot{r}\hat{\mathbf{n}}_r + r\dot{\varphi}\hat{\mathbf{n}}_\varphi.$$

The physical interpretation of this formula is obvious. The first term corresponds to the radial velocity of the particle, and the second one to its angular motion. One can see the illustration for this formula in Fig. 2.7. For the square of the velocity, we have $v^2 = \dot{r}^2 + r^2\dot{\varphi}^2$.

The next step is to calculate the acceleration. By using the relations (2.27), one can write

$$\mathbf{a} = \frac{d\mathbf{v}}{dt} = \frac{d}{dt}(\dot{r}\hat{\mathbf{n}}_r + r\dot{\varphi}\hat{\mathbf{n}}_\varphi) = \ddot{r}\hat{\mathbf{n}}_r + \dot{r}\dot{\varphi}\hat{\mathbf{n}}_\varphi + \dot{r}\dot{\varphi}\hat{\mathbf{n}}_\varphi + r\ddot{\varphi}\hat{\mathbf{n}}_\varphi - r\dot{\varphi}^2\hat{\mathbf{n}}_r$$

$$= \hat{\mathbf{n}}_r(\ddot{r} - r\dot{\varphi}^2) + \hat{\mathbf{n}}_\varphi(2\dot{r}\dot{\varphi} + r\ddot{\varphi}). \tag{2.28}$$

In order to interpret different terms in this expression, we consider different types of movements. The radial part represents the sum of the term containing $\ddot{r}$, which does not depend on the angular motion, and the term that can be identified as centripetal acceleration (pointing to the origin). For the circular motion, this term coincides with the normal acceleration. The second term of the angular part is related to the angular acceleration and it is the same as the one in the formula (2.22). The term of less immediate interpretation is the first one in the angular part. To interpret this term, it is not enough to consider the movement which is purely circular or purely radial, because in both cases this term vanishes. The origin of this term is that the motion of the vector basis is (generally) accelerated. In what follows we will see that this term has connection to the so-called Coriolis force.

Exercise 3. Consider cylindrical coordinates in three-dimensional space,

$$x = r\cos\varphi, \quad y = r\sin\varphi, \quad z = z.$$

Check that, in this case, the formulas for $\hat{\mathbf{n}}_r$, $\hat{\mathbf{n}}_\varphi$ and their derivatives are the same as for the polar coordinates. On the top of that there is a constant basis vector $\hat{\mathbf{n}}_z = \hat{\mathbf{k}}$. Write the formulas to $\mathbf{r}$, $\mathbf{v}$ and $\mathbf{a}$ in cylindrical coordinates.

$$\textbf{Answer:} \quad \mathbf{r} = r\hat{\mathbf{n}}_r + \hat{\mathbf{k}}z;$$

$$\mathbf{v} = \dot{r}\hat{\mathbf{n}}_r + r\dot{\varphi}\hat{\mathbf{n}}_\varphi + \hat{\mathbf{k}}\dot{z};$$

$$\mathbf{a} = \hat{\mathbf{n}}_r(\ddot{r} - r\dot{\varphi}^2) + \hat{\mathbf{n}}_\varphi(2\dot{r}\dot{\varphi} + r\ddot{\varphi}) + \hat{\mathbf{k}}\ddot{z}.$$

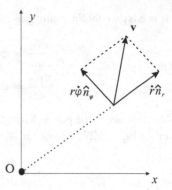

Fig. 2.7 Velocity components
in polar coordinates

As a final example, consider spherical coordinates (r, θ, φ), defined according to Fig. 2.8 and related to the Cartesian coordinates by the equations

$$\begin{cases} x = r\cos\varphi\sin\theta \\ y = r\sin\varphi\sin\theta \\ z = r\cos\theta. \end{cases}$$

Let $\hat{\mathbf{n}}_r$, $\hat{\mathbf{n}}_\varphi$ and $\hat{\mathbf{n}}_\theta$ be the basis vectors indicated in the figure above. The vectors $\hat{\mathbf{n}}_r$ and $\hat{\mathbf{n}}_\theta$ belong to the same vertical plane containing the axis z. Obviously,

$$\hat{\mathbf{n}}_r = \frac{\mathbf{r}}{r} = \frac{x\hat{\mathbf{i}} + y\hat{\mathbf{j}} + z\hat{\mathbf{k}}}{r}$$
$$= \sin\theta\left(\hat{\mathbf{i}}\cos\varphi + \hat{\mathbf{j}}\sin\varphi\right) + \hat{\mathbf{k}}\cos\theta \qquad (2.29)$$

and also

$$\hat{\mathbf{n}}_\varphi = -\hat{\mathbf{i}}\sin\varphi + \hat{\mathbf{j}}\cos\varphi,$$

as in the case of polar coordinates in the plane.

Exercise 4. Verify that

$$\hat{\mathbf{n}}_r \cdot \hat{\mathbf{n}}_r = \hat{\mathbf{n}}_\varphi \cdot \hat{\mathbf{n}}_\varphi = 1 \qquad \text{and} \qquad \hat{\mathbf{n}}_r \cdot \hat{\mathbf{n}}_\varphi = 0.$$

To define the last component of the basis, the vector $\hat{\mathbf{n}}_\theta$, we remember that this should be a unit vector and orthogonal to the other two. So the easiest way is to define this vector as

$$\hat{\mathbf{n}}_\theta = \hat{\mathbf{n}}_\varphi \times \hat{\mathbf{n}}_r. \qquad (2.30)$$

Using the last formula, we have

$$\hat{\mathbf{n}}_\theta = \left(\hat{\mathbf{i}}\cos\varphi + \hat{\mathbf{j}}\sin\varphi\right)\cos\theta - \hat{\mathbf{k}}\sin\theta. \qquad (2.31)$$

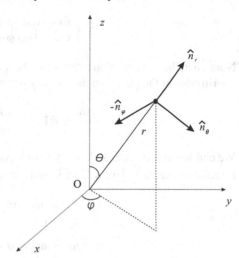

Fig. 2.8 Spherical coordinates

Looking at Fig. 2.8, we can note that the projections of $\hat{\mathbf{n}}_\theta$ and $\hat{\mathbf{n}}_r$ in the plane xy are parallel,[2] and also that the angle between these vectors is $\pi/2$, which can be verified by inspection of (2.29) and (2.31).

Exercise 5. *(a)* Verify that the solution (2.31) satisfies relation (2.30) and $|\hat{\mathbf{n}}_\theta| = 1$.
(b) The property (2.30) means that the set of vectors $(\hat{\mathbf{n}}_r, \hat{\mathbf{n}}_\theta, \hat{\mathbf{n}}_\varphi)$ has dextrorotatory direction, just like the set of vectors $(\hat{\mathbf{i}}, \hat{\mathbf{j}}, \hat{\mathbf{k}})$. Check that the cyclic property is also valid in this case, i.e.,

$$\hat{\mathbf{n}}_r = \hat{\mathbf{n}}_\theta \times \hat{\mathbf{n}}_\varphi, \quad \hat{\mathbf{n}}_\varphi = \hat{\mathbf{n}}_r \times \hat{\mathbf{n}}_\theta.$$

(c) Looking at Fig. 2.8, verify that all the three basis vectors are pointing in the growing direction of the corresponding parameter.

Exercise 6. Check if there are positions of $\mathbf{r}$ for which $\hat{\mathbf{n}}_r = \hat{\mathbf{i}}$ and the other two vectors $\hat{\mathbf{n}}_\theta$ and $\hat{\mathbf{n}}_\varphi$ coincide respectively with (a) $-\hat{\mathbf{k}}$ and $+\hat{\mathbf{j}}$; or with (b) $-\hat{\mathbf{k}}$ and $-\hat{\mathbf{j}}$. Explain why the solution exists in only one of these two cases.

Answer: In the case (b) no solution is possible because the set of vectors $(\hat{\mathbf{i}}, -\hat{\mathbf{k}}, -\hat{\mathbf{j}})$ has different orientation from that of the set of vectors $(\hat{\mathbf{i}}, \hat{\mathbf{j}}, \hat{\mathbf{k}})$.

In order to find the velocity and acceleration in spherical coordinates, we will first calculate

$$\frac{d\hat{\mathbf{n}}_r}{dt}, \quad \frac{d\hat{\mathbf{n}}_\varphi}{dt}, \quad \frac{d\hat{\mathbf{n}}_\theta}{dt}.$$

To simplify the calculation, we introduce the new notations,

---

[2] Without considering the case $\hat{\mathbf{r}} = \pm\hat{\mathbf{k}}$.

$$\begin{cases} \hat{\mathbf{m}} = \hat{\mathbf{i}}\cos\varphi + \hat{\mathbf{j}}\sin\varphi \\ \hat{\mathbf{l}} = -\hat{\mathbf{i}}\sin\varphi + \hat{\mathbf{j}}\cos\varphi. \end{cases}$$

Note that these vectors are the same that we have already found for the polar coordinates in the plane, so they satisfy the relations (2.27),

$$\frac{d\hat{\mathbf{m}}}{dt} = \dot{\varphi}\hat{\mathbf{l}}, \qquad \frac{d\hat{\mathbf{l}}}{dt} = -\dot{\varphi}\hat{\mathbf{m}}. \qquad (2.32)$$

We can see that $\hat{\mathbf{n}}_\varphi$ is exactly the vector $\hat{\mathbf{l}}$. As far as the basis is orthonormal, we get by rotating the basis $\{\hat{\mathbf{m}}, \hat{\mathbf{k}}\}$, the following result:

$$\hat{\mathbf{n}}_r = \hat{\mathbf{m}}\sin\theta + \hat{\mathbf{k}}\cos\theta,$$
$$\hat{\mathbf{n}}_\varphi = \hat{\mathbf{l}},$$
$$\hat{\mathbf{n}}_\theta = \hat{\mathbf{m}}\cos\theta - \hat{\mathbf{k}}\sin\theta. \qquad (2.33)$$

Using formulas (2.33) and (2.32), one can find

$$\frac{d\hat{\mathbf{n}}_\varphi}{dt} = -\hat{\mathbf{m}}\dot{\varphi},$$
$$\frac{d\hat{\mathbf{n}}_\theta}{dt} = \hat{\mathbf{l}}\dot{\varphi}\cos\theta + \dot{\theta}\left(-\hat{\mathbf{m}}\sin\theta - \hat{\mathbf{k}}\cos\theta\right),$$
$$\frac{d\hat{\mathbf{n}}_r}{dt} = \hat{\mathbf{l}}\sin\theta\,\dot{\varphi} + \dot{\theta}\left(\hat{\mathbf{m}}\cos\theta - \hat{\mathbf{k}}\sin\theta\right).$$

At this point, we must invert the relations for $\hat{\mathbf{n}}_\theta$ and $\hat{\mathbf{n}}_r$.

Exercise 7. Check that the relationship between different basis vectors can be written in a matrix form,

$$\begin{pmatrix} \hat{\mathbf{n}}_\theta \\ \hat{\mathbf{n}}_r \end{pmatrix} = \begin{pmatrix} \cos\theta & -\sin\theta \\ \sin\theta & \cos\theta \end{pmatrix} \begin{pmatrix} \hat{\mathbf{m}} \\ \hat{\mathbf{k}} \end{pmatrix}.$$

This is a rotation matrix (See Fig. 2.9) and its inverse matrix has the form

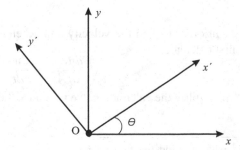

Fig. 2.9 Rotation of a basis

$$\begin{pmatrix} \cos\theta & -\sin\theta \\ \sin\theta & \cos\theta \end{pmatrix}^{-1} = \begin{pmatrix} \cos\theta & \sin\theta \\ -\sin\theta & \cos\theta \end{pmatrix}.$$

Then, the inverse relationship for the basis vectors is given by

$$\begin{pmatrix} \hat{\mathbf{m}} \\ \hat{\mathbf{k}} \end{pmatrix} = \begin{pmatrix} \cos\theta & \sin\theta \\ -\sin\theta & \cos\theta \end{pmatrix} \begin{pmatrix} \hat{\mathbf{n}}_\theta \\ \hat{\mathbf{n}}_r \end{pmatrix},$$

or

$$\hat{\mathbf{m}} = \hat{\mathbf{n}}_\theta \cos\theta + \hat{\mathbf{n}}_r \sin\theta, \qquad \hat{\mathbf{k}} = -\hat{\mathbf{n}}_\theta \sin\theta + \hat{\mathbf{n}}_r \cos\theta.$$

Hence,

$$\frac{d\hat{\mathbf{n}}_\varphi}{dt} = -\dot{\varphi}\left(\hat{\mathbf{n}}_\theta \cos\theta + \hat{\mathbf{n}}_r \sin\theta\right);$$

$$\frac{d\hat{\mathbf{n}}_\theta}{dt} = \dot{\varphi}\hat{\mathbf{n}}_\varphi \cos\theta - \dot{\theta}\hat{\mathbf{n}}_r;$$

$$\frac{d\hat{\mathbf{n}}_r}{dt} = \dot{\varphi}\hat{\mathbf{n}}_\varphi \sin\theta + \dot{\theta}\hat{\mathbf{n}}_\theta. \tag{2.34}$$

Using the last formulas, we calculate the velocity $\mathbf{v}$ and acceleration $\mathbf{a}$ in spherical coordinates. The radius-vector of the particle can be written as

$$\mathbf{r} = r\hat{\mathbf{n}}_r.$$

Taking the time derivative and using (2.34), we have

$$\mathbf{v} = \dot{r}\hat{\mathbf{n}}_r + r\dot{\varphi}\hat{\mathbf{n}}_\varphi \sin\theta + r\dot{\theta}\hat{\mathbf{n}}_\theta. \tag{2.35}$$

An important result is the square of the absolute value of the velocity,

$$\mathbf{v}\cdot\mathbf{v} = v^2 = \dot{r}^2 + r^2\dot{\varphi}^2 \sin^2\theta + r^2\dot{\theta}^2. \tag{2.36}$$

To obtain the acceleration, one needs to do a bit more involved calculation. The result is given by the expression

$$\begin{aligned} \mathbf{a} = \ &\hat{\mathbf{n}}_r(\ddot{r} - r\dot{\theta}^2 - r\dot{\varphi}^2 \sin^2\theta) \\ &+\hat{\mathbf{n}}_\varphi\left(2r\dot{\varphi}\dot{\theta}\cos\theta + 2\dot{\varphi}\dot{r}\sin\theta + r\ddot{\varphi}\sin\theta\right) \\ &+\hat{\mathbf{n}}_\theta(2\dot{r}\dot{\theta} + r\ddot{\theta} - r\dot{\varphi}^2 \sin\theta\cos\theta). \end{aligned} \tag{2.37}$$

Exercises

1. (a) Verify the formulas (2.35), (2.36) and (2.37);
(b) Consider the plane $\theta = \frac{\pi}{2}$ and check the coincidence of the above results with the ones for the polar system in the plane;

(c) Consider $\theta = \theta_0$ and $\varphi = \varphi_0$ (θ_0 and φ_0 constants). Calculate the velocity and acceleration. Give a physical interpretation for the result. What kind of movement is that?

Answer: It is a rectilinear motion, defined by the angles θ_0 and φ_0.

2. Consider the motion of a point mass along the circumference of the radius R, centered at the origin O in the vertical plane $y = 3x$. The particle has a constant angular velocity, ω, and at the initial moment, $t = 0$, has $z = 0$, $\dot{z} > 0$ and $x > 0$. Write components of **r**, **v**, and **a** in (a) Cartesian coordinates, (b) cylindrical coordinates, (c) spherical coordinates.

Answers:

$$\text{(a)} \quad x = \frac{R\cos\omega t}{\sqrt{10}}, \quad y = \frac{3R\cos\omega t}{\sqrt{10}}, \quad z = R\sin\omega t.$$

$$\text{(b)} \quad r = \sqrt{x^2 + y^2} = R|\cos\omega t|, \quad z = R\sin\omega t.$$

$$\text{(c)} \quad r \equiv R, \quad \theta = \frac{\pi}{2} - \omega t.$$

In both cases (b) and (c) we have, for the angular coordinate φ,

$$\varphi = \arctan 3 \quad \text{for} \quad \frac{(2n-1)\pi}{2\omega} \leq t \leq \frac{(2n+1)\pi}{2\omega},$$

$$\varphi = \arctan 3 + \pi \quad \text{for} \quad \frac{(2n+1)\pi}{2\omega} \leq t \leq \frac{(2n+3)\pi}{2\omega}$$

$$\text{where} \quad n = \pm 1, \pm 2, \pm 3, \dots$$

3. Derive the Eq. (2.27) for circular motion using Eq. (2.37).

4. A body is moving along the curve $y = Bx^2$, with $B = const$ and $\ddot{x} = a_0 = const$. Determine the absolute value of the acceleration as a function of x, if $\dot{x}(t = 0) = 0$ and also $x(t = 0) = 0$. Calculate the curvature radius of the trajectory as a function of x.

Answers: $a = a_0\sqrt{1 + 36B^2x^2}$ and $R = \frac{1}{2B}\left(1 + 4B^2x^2\right)^{3/2}$.

5. A material point is moving in a spiral trajectory, according to the equations

$$x = b\cos\omega t, \quad y = b\sin\omega t, \quad z = v_0 t,$$

where b, ω and v_0 are constants. Calculate the radius of curvature of the trajectory at any point and the value of the tangential and normal components of acceleration.

Answers: $a_t = 0$, $a_n = \omega^2 b$ and $R = b + v_0^2/a_n$.

6. A train is moving in a circle of radius $R = 400$ m and its tangential acceleration is 0.2 m/s². Determine the magnitude of its acceleration as well as its normal acceleration at the time instant when its velocity reaches a value of 10 m/s.

Answers: $a_n = 0.25$ m/s² and $a = 0.32$ m/s².

7. Knowing that the velocity of a particle is given by

$$\mathbf{v} = 4\omega R\hat{\mathbf{i}}\cos\omega t + 4\omega R\hat{\mathbf{j}}\sin\omega t + 3\omega R\hat{\mathbf{k}}$$

and the coordinates at the initial moment are $x(0) = R$, $y(0) = -4R$ and $z(0) = 0$, calculate the position vector, $\mathbf{r}(t)$, and the acceleration, $\mathbf{a}(t)$. Get the formulas for the normal and tangential components of acceleration. Compare the results with the one in Exercise 5.

8. Calculate the normal and tangential components of acceleration of a particle that is moving in a homogeneous gravitational field, $\mathbf{g}$. The initial velocity, $\mathbf{v}_0$, makes an angle α with the vector $\mathbf{g}$. Determine the radius of curvature of this trajectory as a function of time.

$$\textbf{Answer:}\quad R = \frac{\left(v_0^2 + g^2 t^2 + 2v_0 g t \cos\alpha\right)^{3/2}}{v_0 g \sin\alpha}.$$

9. A particle moves so that its polar coordinates are given by $\rho(t) = v_0 t$ and $\phi(t) = \omega_0 t$, where v_0 and ω_0 are positive constants. Sketch the path of the particle. Write $\mathbf{v}$ and $\mathbf{a}$ in the polar basis with its constituents as a function of time. Calculate the instantaneous radius of curvature of the trajectory, R in terms of ϕ. After a large number of turns, N, calculate R.

Solution. We can write $\mathbf{v} = \dot{\rho}\hat{\rho} + \rho\dot{\phi}\hat{\phi} = v_0\hat{\rho} + v_0\omega_0 t\hat{\phi}$. Taking the derivative, we obtain $\mathbf{a} = 2\omega_0 v_0\hat{\phi} - v_0\omega_0^2 t\hat{\rho}$. To calculate R, just take the modulus of the equation

$$2\omega_0 v_0\hat{\phi} - v_0\omega_0^2 t\hat{\rho} = \frac{dv}{dt}\hat{v} + \frac{v^2}{R}\hat{n}.$$

Using $v = v_0\sqrt{1 + \phi^2}$, then we have

$$R(\phi) = \frac{v_0}{\omega_0}\frac{(\phi^2 + 1)^{3/2}}{\phi^2 + 2}.$$

For $\phi = 2\pi N \gg 1$, we have $R \to v_0\phi/\omega_0 = 2\pi N v_0/\omega_0$.

10. The problem of the fox. A fox is running along a straight line with a constant velocity v, trying to escape from the dog that always runs in the direction of the fox with constant velocity u. At the initial time instant $t = 0$ the dog is at the distance L from the fox and the line connecting both of them is orthogonal to the path of the fox.

(a) Calculate the relationship between the distance that separates the fox from the dog, l, and the angle φ between their velocity vectors;
(b) Determine the acceleration of the dog as a function of the same angle, φ.
(c) For the case $v > u$, calculate the time required for the dog to reach the fox.

Solution. The angle between the vector with components l_x, l_y joining the positions of the animals and the line of motion of the fox is φ. We have $\varphi = \varphi(t)$ and $l = l(t)$ and always

$$l_x = l \cos\varphi, \qquad l_y = \sin\varphi.$$

For an infinitesimal interval of time t, we obviously have

$$dl_x = (u - v\cos\varphi)\,dt, \qquad dl_y = -v\sin\varphi\,dt,$$

so that, using $l\,dl = l_x\,dl_x + l_y\,dl_y$, we have

$$dl = (u\cos\varphi - v)\,dt. \tag{2.38}$$

On the other hand, considering only the terms of first order in the differentials, we obtain

$$l_x + dl_x = (l + dl)\cos(\varphi + d\varphi) = l_x + \cos\varphi\,dl - l\sin\varphi d\varphi,$$

$$l_y + dl_y = (l + dl)\sin(\varphi + d\varphi) = l_y + \sin\varphi\,dl + l\cos\varphi d\varphi.$$

Using these relations, we can write

$$dl_x = \cos\varphi\,dl - l\sin\varphi\,d\varphi, \qquad dl_y = \sin\varphi\,dl + l\cos\varphi\,d\varphi.$$

Inverting the above relations, we obtain

$$dl = \cos\varphi\,dl_x + \sin\varphi\,dl_y, \qquad l\,d\varphi = -\sin\varphi\,dl_x + \cos\varphi\,dl_y.$$

As a result, we conclude that

$$l\,d\varphi = -u\sin\varphi\,dt. \tag{2.39}$$

Using the relations (2.38) and (2.39), we can eliminate time and arrive at the differential equation

$$\frac{l\,d\varphi}{dl} = \frac{u\sin\varphi}{v - u\cos\varphi}.$$

This equation is separable and its solution can be written as

$$\frac{l(\varphi)}{C} = (1 - \cos\varphi)^{\frac{v-u}{2u}}\,(1 + \cos\varphi)^{-\frac{v+u}{2u}}, \tag{2.40}$$

where C is a constant of integration. Using the initial condition $l(\varphi = \pi/2) = L$, the constant is defined by $C = L$. It is easy to see that the dog reaches the fox in the case $v > u$, while in the case $v \le u$ the fox will escape from the dog.

To find the acceleration of the dog, first we note that its tangential component is zero, because the module of its velocity is constant. To determine the normal component of acceleration, one can use (2.39) and obtain

$$a_n = \dot{\varphi} v = -\frac{u \sin \varphi}{l(\varphi)},$$

where the function $l(\varphi)$ is given by Eq. (2.40) with $C = L$.

Finally, to calculate the time of the race, T in the case $u > v$ (otherwise, T is infinite) we note that, according to the result (2.38),

$$L = \int_0^T (u \cos \varphi - v) \, dt. \tag{2.41}$$

On the other hand, by the time T, the projections of the positions of the two animals on the line that defines the trajectory of the fox suffers the same increment, i.e.,

$$\int_0^T v \cos \varphi \, dt = uT. \tag{2.42}$$

Comparing the Eqs. (2.41) and (2.42), we obtain the result

$$T = \frac{vL}{u^2 - v^2}.$$

2.4 Kinematics of a Rigid Body

In this section, we consider the kinematics of the rigid body. This part of mechanics has a special importance because of its applications, apart from the fact that in many cases it is useful to associate a reference frame with a rigid body. Sometimes it is useful, for example, to work in a rotating reference frame, and find its relation to the inertial (laboratory) reference frame.

To study the motion of a rigid body, it proves useful to introduce a static reference frame with the origin O_o (we will identify it as the laboratory reference frame), and also another, non-inertial reference system attached to the rigid body, with a mobile origin, O, as shown in Fig. 2.10. It is not necessary that the point O belongs to the body, but the distances between this point and all the points of the body must be fixed.

Any movement of the rigid body can be considered as the sum of two independent motions. The first one is the movement of the point O in relation to the reference frame of the laboratory, point O_o. This movement is nothing else, but the motion of a point which can be parameterized by the radius-vector $\mathbf{r}_0(t)$. The second movement is the one of the rigid body with a static (fixed) point O. For a point M of the body, this movement can be parameterized by the radius-vector $\mathbf{r}(t)$. Then we have $\mathbf{R}(t) = \mathbf{r}_0(t) + \mathbf{r}(t)$.

It is clear that the point O can perform complicated motions with respect to O_o but anyway, it is a motion of a single point. For its description we can use the formalism developed in the previous chapter. So, in order to get qualitatively new information,

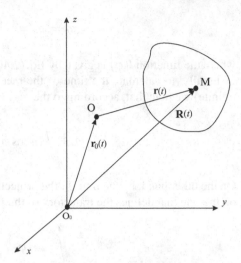

Fig. 2.10 Rigid body in the co-moving reference system and at the laboratory reference frame

it is sufficient to consider the reference frame where O is a static point. In this case, the movement of the rigid body is reduced to a sequence of rotations around an instantaneous axis.

A complete description of the motion of a rigid body, with the variable directions of the axis can only be performed in the context of analytical mechanics. In this book we are going to consider the motion around a instantaneous axis in a simplified form. We will mainly consider the motion of bodies with fixed axis or with a relatively simple geometry and mass distribution.

The first step is to study the rotation of a rigid body around a fixed axis passing through the point O (see Fig. 2.11).

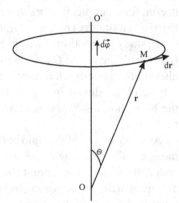

Fig. 2.11 Rotation of a point of the rigid body, M, around the axis OO'

When the rotation axis is fixed, we can focus the attention on the angle of rotation, φ. We adopt the *radian* as the unit of angular measure. For an infinitesimal rotation, we can associate an angle of rotation with the vector $d\vec{\varphi}$, such that

$d\vec{\varphi} \parallel OO'$ and it has the direction defined by the right-hand rule (or rule of the corkscrew). For a point M, with its position specified by the radius-vector $\mathbf{r}$, its displacement due to $d\vec{\varphi}$ is given by

$$dr = d\vec{\varphi} \times \mathbf{r} = \left[d\vec{\varphi}, \mathbf{r} \right], \tag{2.43}$$

where we have used two different notations for the vector product (see Mathematical Appendix).

An important observation is related to the possibility of reversing the coordinate axes (this transformation is called space inversion, or parity inversion),

$$x \rightarrow -x, \qquad y \rightarrow -y, \qquad z \rightarrow -z. \tag{2.44}$$

What is the behavior of the components of the vectors $\mathbf{r}$, $d\mathbf{r}$, and $d\vec{\varphi}$ in this case? Obviously, for any radius-vector, including $\mathbf{r}$ and $\mathbf{r} + d\mathbf{r}$, the components change their signs. In particular, besides (2.44), we have $x + dx \rightarrow -(x + dx)$ etc. We can describe this situation as the transformation $\mathbf{r} \rightarrow -\mathbf{r}$, in a sense that all components of the vector change their signs. At the same time $\mathbf{r} + d\mathbf{r} \rightarrow -(\mathbf{r} + d\mathbf{r})$, that means $d\mathbf{r} \rightarrow -d\mathbf{r}$. In this case, in order to maintain the orientation of the vectors $\mathbf{r}$, $d\vec{\varphi}$, $d\mathbf{r}$ and the relation (2.43), we need to define the rule of transformation of the vector $d\vec{\varphi}$ to be different, namely we need it to transform without change of sign,

$$d\vec{\varphi} \rightarrow d\vec{\varphi}. \tag{2.45}$$

In other words, in order to keep the consistency of the formulas, the components of the vector $d\vec{\varphi}$ should not suffer a change when the basis vectors turn to the opposite sides and the components of the radius-vector change their signs. We conclude that there are two different kinds of vectors:

1) Simple vectors, such as $\mathbf{r}$, $\mathbf{v}$ and $\mathbf{a}$. Under the inversion of axes their components change their signs. In this sense we can say that the vectors change their directions to the opposite ones, e.g., $\mathbf{r} \rightarrow -\mathbf{r}$, $\mathbf{v} \rightarrow -\mathbf{v}$, etc.

2) Axial vectors, such as $d\vec{\varphi}$. They do not change the signs of their components (one can say that they do not change their directions, in this sense) under the inversion of the axes.

Both types of vectors have important applications in physics. Typically, axial vectors appear as a consequence of the cross product. For example, the position, velocity and acceleration of a particle are vectors (simple vectors). But if we recall the formula for the Lorentz force exerted by the magnetic field, $\mathbf{H}$, on a charged particle moving with velocity $\mathbf{v}$,

$$\mathbf{F} = m\mathbf{a} = e\mathbf{v} \times \mathbf{H},$$

it becomes clear that the magnetic field must be an axial vector.

Dividing $d\vec{\varphi}$ by dt, we find the angular velocity $\vec{\omega} = d\vec{\varphi}/dt$. In the general case, the direction of $\vec{\omega}$ depends on time. This means that the axis of rotation may depend on time as well as the magnitude of $\vec{\omega}$. The derivative of $\vec{\omega}$ is the angular acceleration vector, $\vec{\beta} = d\vec{\omega}/dt$. The units for $\vec{\omega}$ and $\vec{\beta}$ are rad/s and rad/s^2 respectively.

Now we can find a relationship between linear and angular quantities. Dividing (2.43) by dt, we obtain

$$\frac{d\mathbf{r}}{dt} = \frac{d\vec{\phi}}{dt} \times \mathbf{r}, \qquad \text{i.e.,} \qquad \mathbf{v} = \vec{\omega} \times \mathbf{r}. \tag{2.46}$$

The last formula is valid for the rotations of a rigid body relative to the axis OO' that passes through a static point, O. In order to find the analogous relation for the acceleration, one has to take the derivative of the last relationship (see Mathematical Appendix for the derivative of the vector product):

$$\frac{d\mathbf{v}}{dt} = \frac{d\vec{\omega}}{dt} \times \mathbf{r} + \vec{\omega} \times \frac{d\mathbf{r}}{dt} = \vec{\beta} \times \mathbf{r} + \vec{\omega} \times \mathbf{v}.$$

Using (2.46), we obtain

$$\mathbf{a} = [\vec{\beta}, \mathbf{r}] + [\vec{\omega}, [\vec{\omega}, \mathbf{r}]].$$

The last term can be rewritten using the formula for the double vector product,

$$[\mathbf{a}, [\mathbf{b}, \mathbf{c}]] = \mathbf{b}(\mathbf{a}, \mathbf{c}) - \mathbf{c}(\mathbf{a}, \mathbf{b}), \tag{2.47}$$

and so we arrive at the general expression for the acceleration,

$$\mathbf{a} = \vec{\beta} \times \mathbf{r} + \vec{\omega}(\vec{\omega} \cdot \mathbf{r}) - \mathbf{r}\omega^2. \tag{2.48}$$

The last formula becomes simpler when we consider a particle moving on the plane which includes the origin, point O, and is perpendicular to the rotation axis. In this case, $\mathbf{r} \perp \vec{\omega}$ and the expression (2.48) reduces to

$$\mathbf{a} = \vec{\beta} \times \mathbf{r} - \mathbf{r}\omega^2. \tag{2.49}$$

In the particular case when the axis of rotation does not depend on time, the direction of the angular velocity is constant, i.e., $\vec{\beta} \parallel \vec{\omega}$ and we can see that the two terms in (2.49) are vectors with orthogonal directions. The first term, $\vec{\beta} \times \mathbf{r}$, always has the direction perpendicular to the radius-vector, and the second term, $-\mathbf{r}\omega^2$, is the usual expression for the centripetal acceleration.

Finally, we write the formulas for the motion of the point M of the rigid body (see Fig. 2.11) in the reference of the laboratory, i.e., referred to the point O_o (starting point of the static coordinate system). As we know, the radius-vector in the laboratory reference system is given by

$$\mathbf{R}(t) = \mathbf{r}_0(t) + \mathbf{r}(t).$$

Taking the derivative of both sides, we find

$$V(t) = \frac{dR}{dt} = v_0(t) + v(t),$$

where $v_0(t)$ is the velocity of the point O related to the point O_o. Using the previous results, we obtain

$$V(t) = v_0(t) + \vec{\omega} \times r = v_0(t) + \vec{\omega} \times (R - r_0).$$

The last form represents a relationship between the radius-vector R and its derivative V.

For the acceleration we can use the relation (2.48) and write a similar formula

$$a = a_0 + \vec{\beta} \times r + \vec{\omega}(\vec{\omega} \cdot r) - r\omega^2,$$

where

$$a_0 = \frac{dv_0}{dt} = \frac{d^2 r_0}{dt^2} \quad \text{and} \quad r = R - r_0.$$

Exercise 8. (a) Verify the formula (2.47), using the definition of vector product

$$b \times c = [b, c] = \begin{vmatrix} \hat{i} & \hat{j} & \hat{k} \\ b_x & b_y & b_z \\ c_x & c_y & c_z \end{vmatrix}.$$

(b) Consider the case where $\vec{r} \parallel \vec{\omega}$ and interpret the result for v and a.

Exercise 9. (a) Consider the vector quantities listed below and identify which ones are vectors (simple vectors) and which are axial vectors.

(i) r, (ii) v, (iii) a,
(iv) $l_1 = r \times v$, (v) $l_2 = r \times a$, (vi) $l_3 = v \times a$,
(vii) $l_4 = r \times l_1$, $(viii)$ $l_5 = r \times l_3$.

(b) Try to get the same answers using different approaches.
(c) Try to elaborate a general criteria for identifying the nature of the vectors with respect to inversion of the coordinates, namely whether a given vector is a simple one or an axial vector.
(d) Check the result for the vector l_4 using the Eq. (2.47).

Exercise 10. Consider the motion of a rigid body with constant rotation axis which coincides with the axis OX. Compare the velocities v_1 and v_2 of the points of the radius-vectors

$$r_1 = 3\hat{i} + 2\hat{j} + \hat{k} \quad \text{and} \quad r_2 = 5\hat{i} + \hat{j} + 2\hat{k}.$$

(a) What is the ratio between the magnitudes of these velocities?
(b) Why does one of the components of each **r** have no importance for the answer?
(c) Find a rotation axis that passes through the point O, for which the magnitudes of the velocities are equal. Is this the unique solution for this problem?

Exercise 11. Considering the previous problem, assume that the angular velocity is given by $\vec{\omega} = \omega\hat{\mathbf{i}}$, with $\omega = const$, and calculate the instantaneous accelerations $\mathbf{a}_1$ and $\mathbf{a}_2$ for the points of the rigid body with radius vectors $\mathbf{r}_1$ and $\mathbf{r}_2$.

2.5 Transformation Between Reference Systems

In this section, we consider the transformation of coordinates, velocities and accelerations between different reference frames. Remember that the reference frame is a spatial coordinate system defined by an origin O, and by the linearly independent vectors $\hat{\mathbf{n}}_1$, $\hat{\mathbf{n}}_2$, and $\hat{\mathbf{n}}_3$. Any position vector which starts at the origin O and ends at the given point, can be written as

$$\mathbf{r} = x^1\hat{\mathbf{n}}_1 + x^2\hat{\mathbf{n}}_2 + x^3\hat{\mathbf{n}}_3,$$

where x^1, x^2 and x^3 are coordinates in the given reference frame. Our goal is to accomplish the transformation of the components of the vectors **r**, **v**, and **a** from one reference frame to another. In this book, we consider only the cases when the basis is orthonormal, i.e., $\hat{\mathbf{n}}_i \cdot \hat{\mathbf{n}}_j = \delta_{ij}$. Here we introduced a useful notion of the Kronecker delta δ_{ij} (or Kronecker symbol) which is equal to 1 when $i = j$ and to 0 when $i \neq j$.

First we will study the cases when, for both reference frames, the basis vectors are the same, namely: $\hat{\mathbf{i}}, \hat{\mathbf{j}}$, and $\hat{\mathbf{k}}$; and only the initial points O and O' are different. Under these conditions, the relationship between the reference points can be described as a transformation called *translation*, or *displacement*. Suppose that $\overrightarrow{OO'} = \mathbf{r}_0$. For a spatial point M, we have in this case (see Fig. 2.12),

$$\mathbf{r} = \mathbf{r}' + \mathbf{r}_0.$$

At this point we can introduce the notion of inertial reference frame, which is very important in mechanics. For example, Newton's first law establishes the existence of an inertial reference frame. When the coordinate system K is inertial, any free particle (i.e., free of interactions with other objects) remains in motion with constant velocity, $\mathbf{v} = const$. In practice, it is impossible to find a physical body that has no interaction with any other object. We can isolate an object, for example, taking him into space, far from the Earth. But he will still be under the action of the gravitational field of the Sun. Taking the same object away from the Solar system, there is still a galaxy's gravitational field and so on. A free body is an idealization.

A possible option to achieve inertial reference frame in reality is to let the body to move freely in a weak gravitational field which should be approximately

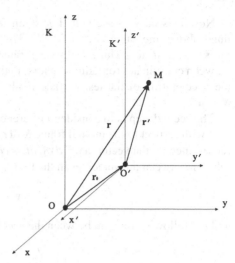

Fig. 2.12 Two reference
frames related by a translation

homogeneous. In this case, the gravitational force effectively disappears for this
body, and we can consider the reference frame attached to it as being inertial.
We will discuss this issue later in the chapters on the laws of dynamics, when considering gravity.

In Mechanics, we can make a description of movements both in an inertial
reference frame, and in a non-inertial one. In the present chapter, dedicated to kinematics, we will start by considering the coordinate transformations between inertial
reference frames, and after that we will consider transformations to non-inertial reference frames.

Suppose K is an inertial reference frame, which can be considered static, for
simplicity. Another reference frame, K', is moving with respect to K with constant
velocity, $\mathbf{v}_0$. The relationship between the radius-vectors $\mathbf{r}$ and $\mathbf{r}'$ of the same point
M, in both reference frames is

$$\mathbf{r} = \mathbf{r}' + \mathbf{v}_0 \cdot t. \tag{2.50}$$

The last formula is called Galileo's transformation. If the relative velocity (velocity
of K' with respect to K) is directed along the OX axis to the positive side, i.e.,
$\mathbf{v}_0 = v_0\hat{\mathbf{i}}$, we will have the following relations between the two sets of coordinates:

$$\begin{aligned}
x &= x' + v_0 t, \\
y &= y', \\
z &= z', \\
t &= t'.
\end{aligned} \tag{2.51}$$

In order to obtain the velocities of the particles in the two reference frames, we take
the time derivative of (2.50) and obtain

$$\mathbf{v} = \mathbf{v}' + \mathbf{v}_0.$$

Now it is easy to see that if K is an inertial reference frame in the way defined above, the same is true for K'. The reason is that if $\mathbf{v} = const$, then $\mathbf{v}' = \mathbf{v} - \mathbf{v}_0 = const$ too. Thus, the Galileo transformation (2.51) shows the relationship between coordinates (or radius-vectors) in the two inertial frames. Since $\mathbf{v}_0 = const$, the acceleration of the test particle is always the same in two reference frames, $\mathbf{a}' = \mathbf{a}$.

The second step is to consider a reference frame K', which is uniformly accelerated with respect to the inertial frame K. The acceleration of K' with respect to K is represented by the vector $\mathbf{a}_0 = d\mathbf{v}_0/dt = const$. In this case, we can easily find the relation between the velocities in the two frames,

$$\mathbf{v}' = \mathbf{v} - \mathbf{a}_0 t \tag{2.52}$$

and the following relation between the accelerations in the two reference frames:

$$\mathbf{a}' = \mathbf{a} - \mathbf{a}_0. \tag{2.53}$$

For example, if the material point has constant velocity, $\mathbf{v} = const$, in the frame K, the same point will move with acceleration $\mathbf{a}' = -\mathbf{a}_0$ with respect to K'. According to the formula (2.52), K' is not an inertial frame, because even if no force acts on this body, its velocity $\mathbf{v}'$ is not a constant.

The third step is to consider the case where K is an inertial reference frame and K' is spinning around a fixed axis. To simplify the consideration, let us choose the axis OZ to be the axis of rotation, and assume that the origins of the two coordinate systems coincide, $O = O'$ (see Fig. 2.13). In this case, we have $\vec{\omega} = -\omega\hat{\mathbf{k}}$ and can use the results concerning rotation of the rigid body from the previous chapter. The radius-vector of the point M in both frames is the same at every instant, i.e.,

$$\mathbf{r}' = \overrightarrow{O'M} = \overrightarrow{OM} = \mathbf{r}. \tag{2.54}$$

At the same time, the velocities and accelerations in the two reference frames differ from each other, as we shall see.

Let us suppose that the point M is moving with the speed $\mathbf{u}'$ with respect to the reference frame K', at the moment when its radius-vector is given by $\mathbf{r} = \mathbf{r}'$. Consider a very small (infinitesimal) time interval Δt. The displacement of the point M with respect to the rotating reference frame K' is $\Delta\mathbf{r}' = \mathbf{u}'\Delta t$. At the same time, the displacement of a point which is fixed in the reference frame K' at the same point M, with respect to K is $\mathbf{v}\Delta t = \vec{\omega} \times \mathbf{r}\Delta t$. Thus, the displacement of the point M with respect to static reference frame K will be given by the sum of the two terms,

$$\Delta\mathbf{r} = \left(\mathbf{u}' + \vec{\omega} \times \mathbf{r}\right)\Delta t.$$

Dividing both sides by Δt and taking the limit $\Delta t \to 0$, we observe that the velocity of this point in K is

$$\mathbf{u} = \mathbf{u}' + \vec{\omega} \times \mathbf{r}. \tag{2.55}$$

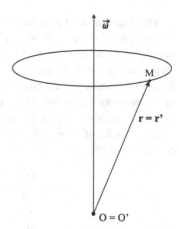

Fig. 2.13 The non-inertial reference frame K' is rotated relative to the inertial one K. The starting points are the same, so $\mathbf{r} = \mathbf{r}'$

The result (2.55) is very interesting because it can be seen in other way. Let us remember, once more, that the position of the vectors M in both frames, K' and K, coincide according to (2.54). Of course, time runs identically in both frames, as always in Classical Mechanics. Apparently, this yields to a contradiction to the relation (2.55) for velocities.

To explain what is the origin for the relation (2.55), one has to understand the main difference between the reference frames K and K'. The frame K is inertial and for the time interval Δt, the displacement of the particle M is given by $\Delta \mathbf{r}$, whereas in respect to K', it is given by $\Delta \mathbf{r}'$. So, despite the coincidence of the radius vectors, their variations are different, because the reference system K' is rotating, i.e., performs its own motion, additional to the one of the motion of the particle in the rotating frame. This additional movement of K' produces an extra term in (2.55). The procedure that generates this term can be seen as a more sophisticated rule to take the time derivative of any vector at the reference frame K', because all vectors which are static in K' are rotating with respect to K with the same angular speed. We can define a "generalized" derivative (called *covariant derivative* by mathematicians) of a vector $\mathbf{b}$, taking into account the rotation of this vector,

$$\frac{D\mathbf{b}}{dt} = \frac{d\mathbf{b}}{dt} - \vec{\omega} \times \mathbf{b}. \qquad (2.56)$$

In this case, we rewrite (2.55) as follows:

$$\frac{D\mathbf{r}'}{dt} = \mathbf{u}' = \frac{d\mathbf{r}'}{dt} - \vec{\omega} \times \mathbf{r}' = \frac{d\mathbf{r}}{dt} - \vec{\omega} \times \mathbf{r}, \qquad (2.57)$$

where we used, once more, the relationship (2.54). For the radius-vector, the "true" time derivative in the reference frame K' is given by D/dt from the Eq. (2.57), and not by the simple derivative d/dt. The additional term appears because of the rotation, which produces an extra displacement of the radius-vector.

The fact that we know how to take the time derivative of an arbitrary vector in a non-inertial rotating frame K' turns out to be very helpful in calculating the acceleration in this frame. An important observation is that (as we have already mentioned before) all vectors which are static in K' are rotating with respect to K with the same angular speed. As a consequence, the rule for taking a derivative in the rotating frame should be the same for all kind of vectors. By applying the covariant derivative in the reference system K' according to (2.56), to the velocity of the particle, $\mathbf{u}' = \mathbf{u} - \vec{\omega} \times \mathbf{r}'$, and remembering that the angular velocity, $\vec{\omega}$, is constant, we obtain

$$
\begin{aligned}
\mathbf{a}' &= \frac{D\mathbf{u}'}{dt} = \frac{d\mathbf{u}'}{dt} - \vec{\omega} \times \mathbf{u}' \\
&= \frac{d}{dt}\left(\mathbf{u} - \vec{\omega} \times \mathbf{r}\right) - \vec{\omega} \times \left(\mathbf{u} - \vec{\omega} \times \mathbf{r}\right) \\
&= \mathbf{a} - \vec{\omega} \times \mathbf{u} - \vec{\omega} \times \mathbf{u} + \vec{\omega} \times \left(\vec{\omega} \times \mathbf{r}\right) \\
&= \mathbf{a} - 2\vec{\omega} \times \mathbf{u} + \vec{\omega} \times \left(\vec{\omega} \times \mathbf{r}\right).
\end{aligned}
$$

In order to obtain the inverse relation between accelerations, one can simply use the formula $\mathbf{r} = \mathbf{r}'$. As a result, we arrive at the expression for acceleration in the frame K,

$$
\begin{aligned}
\mathbf{a} &= \mathbf{a}' + 2\vec{\omega} \times \mathbf{u} - \vec{\omega} \times \left(\vec{\omega} \times \mathbf{r}\right) \\
&= \mathbf{a}' + 2\vec{\omega} \times \mathbf{u}' + 2\vec{\omega} \times \left(\vec{\omega} \times \mathbf{r}'\right) - \vec{\omega} \times \left(\vec{\omega} \times \mathbf{r}'\right) \\
&= \mathbf{a}' + 2\vec{\omega} \times \mathbf{u}' + \vec{\omega} \times \left(\vec{\omega} \times \mathbf{r}'\right).
\end{aligned}
\tag{2.58}
$$

Exercises

1. Write the equations similar to Eq. (2.51) for the cases when the relative velocity between the two frames is directed along the OY and OZ axis and for the general case, when the orientation is arbitrary, e.g.,

$$
\mathbf{v}_0 = v_0 \left(\hat{\mathbf{i}}\cos\alpha + \hat{\mathbf{j}}\cos\beta + \hat{\mathbf{k}}\cos\gamma\right),
$$

where α, β, and γ are the angles between $\mathbf{v}_0$ and the vectors $\hat{\mathbf{i}}, \hat{\mathbf{j}}$, and $\hat{\mathbf{k}}$, respectively.
Answer: In the general case

$$
x = x' + v_0 t \cos\alpha, \quad y = y' + v_0 t \cos\beta, \quad z = z' + v_0 t \cos\gamma, \quad t = t'.
$$

2. A train began to accelerate in the straight part of its path with acceleration a. A small lead block, inside the train, begin to slide horizontally without friction. Determine the acceleration of this block with respect to (a) the Earth; (b) the train.

Write the expression for $v(t)$ of the block in relation to the train.
Answers: *(a)* zero. *(b)* $-a$, $v = -at$.

3. Consider a particular case of the Eq. (2.57),

$$\mathbf{u}' = \frac{D\mathbf{r}'}{dt} = \mathbf{u} - \vec{\omega} \times \mathbf{r},$$

in the case where the velocity $\mathbf{u}'$ is zero. (a) What is the physical interpretation of $\mathbf{u}$ in this case? What is the origin of $\mathbf{u}$? (b) What is the physical result of the condition $\mathbf{u} = 0$? Suggest an example of a system where such a movement occurs.
Answers: *(a)* The condition $\mathbf{u}' = 0$ means that the point is rotating in relation to the reference K and is at rest relative to the reference K'. The velocity $\mathbf{u}$ is only caused by rotation.

(b) The condition $\mathbf{u} = 0$ means that the point is at rest with respect to the inertial reference frame K. Such a point is not rotating together with K'.

Chapter 3
Newton's Laws

Abstract The three Newton's laws form the basis of Classical Mechanics of point-like particles and bodies. We shall present important notions such as the concept of inertial system and forces, and discuss, at the qualitative level, the difference between fundamental and effective interactions and the limits of applicability of the Classical Mechanics. Another important subject is related to the notion of inertia force, coming from consideration of non-inertial reference frames. Special attention will be paid to the Equivalence Principle (which is a key notion in General Relativity), turning out to be a useful tool in solving some problems of Classical Mechanics.

3.1 Introductory Remarks

In the previous chapter we were concerned only with the kinematical description of movements of a material point or a rigid body. Starting from the present chapter we begin to study the dynamics of particles and rigid bodies. The aim of dynamics is to study the origins of the motions. The laws of dynamics allow us to link velocities, positions and accelerations of bodies by relations called dynamical equations. These relations have the form of differential equations for the coordinates of the moving bodies. The equations describing the motion arise as solutions of these equations. These solutions enable one, in principle, to establish the positions of all elements of a mechanical system as functions of time and the positions and velocities of the same elements at the starting instant of time (those are called initial data).

For the formulation of Newton's Laws, we must introduce two new notions: *mass* and *force*. The force, **F**, is a vector quantity that indicates the intensity of the interaction between the two bodies. There are different types of forces for different energy and distance scales. For example, for macroscopic bodies, we can indicate the gravitational force, the Coulomb force between objects with electric charge, the magnetic force that acts between moving charges, the force of elasticity, the force of friction, the force of resistance of air or water and the force of inertia that appears in non-inertial frame of reference. In the next section we will discuss

I.L. Shapiro and G. de Berredo-Peixoto, *Lecture Notes on Newtonian Mechanics*,
Undergraduate Lecture Notes in Physics, DOI 10.1007/978-1-4614-7825-6_3,
© Springer Science+Business Media, LLC 2013

more details about forces. We have to mention, to complete the story, a more general formulation, called Analytical Mechanics, where the force is not the main object in the description of the movements. In Analytical Mechanics, the force depends on other concepts which are regarded to be more fundamental. This description is more natural, in particular, because it is closely related to the main properties of space and time. However, in the present book we will not consider Analytical Mechanics and therefore the concept of force is the fundamental one for us.

The mass of a body is considered as its fundamental feature. A very important property of mass in Classical Mechanics is that it is an additive quantity. If a body is composed of some parts, the mass of the whole is equal to sum of the masses of the parts. An important observation is that this property is not necessary valid for, e.g., elementary particles, where Classical Mechanics should be replaced by other theories, such as Relativistic Mechanics and Quantum Theory. For example, the nucleons (protons and neutrons) are made out of quarks. However, masses of the quarks are responsible only for about 5 % of the mass of the nucleon, the rest is due to the quantum effects. For this reason, the interaction of quarks can not be dealt with in the framework of Classical Mechanics, which has different area of application.

3.2 Inertial Frame of Reference: Newton's First Law

As a first step, one has to choose an appropriate reference frame for the formulation of the laws of dynamics. The simplest choice is the inertial reference frame. As we have previously defined, in this reference frame, a free particle, which is not subject of interaction, remains in uniform motion. This means that the velocity of the particle is constant both in magnitude and in direction. In fact, as it was discussed earlier, no one has ever observed a completely free particle, so the existence of an inertial frame is actually a non-trivial fact. The possibility of such a reference frame is one of the postulates of Classical Mechanics, called *Galileo's Principle* or *Newton's First Law*.

Newton's First Law postulated the existence of an inertial frame. One inertial frame moves in relation to another inertial frame with a constant velocity and hence the law of transformation of the particle's radius-vector is given by

$$\mathbf{r}' = \mathbf{r} + \mathbf{v}t.$$

In any inertial frame, any point in space has the same properties (homogeneity of space), and there is no privileged direction (isotropy of space). According to Newton's first law, all inertial frames are equivalent. This means that all the laws of mechanics are the same in these inertial reference frame. No experience can show a difference between laws of Physics in a static inertial frame and in other inertial frame which is moving with constant velocity with respect to the first one.

Is it possible in practice to find an inertial frame? Let us add some extra discussion to the previous one of the same subject in Sect. 2.5. In some cases, we can consider the Earth itself as an inertial frame, but this is only an approximation. For example, the famous experiment with the Foucault pendulum shows that this is not correct. The complete understanding of the subject of inertial frames is only possible in the context of General Relativity, which states that the forces of inertia and gravitation are locally equivalent. This means that gravity can be eliminated in small volumes if we choose a frame in free fall. But what is the role of other forces in this case? Can we get rid of them?

Let us remember that in modern physics, we only know four kinds of fundamental interactions: electromagnetic, weak, strong and gravitational. The weak and strong forces manifest themselves only at the very small distance scale, e.g., in the nucleus of atoms or in special devices such as particle accelerators, when quantum effects are relevant. In other words, these forces can not be observed at macroscopic distances and the physical conditions in which they are significant are out of the domain of Classical Mechanics. The electromagnetic force can be relevant in mechanics, but it can be eliminated – just consider a body with no charge. All in all, we can see that all other fundamental forces except gravity can not, in principle, prevent the realization of the idea of a free particle.

To better understand the issue of the existence of an inertial frame, we note that there are two kinds of forces, namely inertial and gravitational ones, which are proportional to the mass of the particle. So far, all the experiments have confirmed the equality of the gravitational and inertial masses with great precision. In this respect these two forces are different from other fundamental forces, which are related to specific charges (e.g., electric charge in case of electromagnetic interaction). One can define the ratio between the masses m_1 and m_2 as the ratio of the magnitudes of external gravitational forces F_1^g and F_2^g, acting on these masses, and write $\dfrac{m_1}{m_2} = \dfrac{F_1^g}{F_2^g}$.

The local equivalence between the inertia and gravitation forces means that if we let a very small box fall freely in the gravitational field, we can consider the reference frame attached to the box as an inertial frame. This principle of General Relativity depends on the equality between inertial mass and gravitational mass of a body. In what follows we will see that the use of equivalence principle is a very useful instrument for solving certain problems in the Newtonian Mechanics. In many cases the use of this instrument greatly simplifies the solutions of the problems and also, helps to better understand physical situations.

3.3 Newton's Second Law

Now let us consider **Newton's Second Law**, whereby the acceleration **a** of a body with mass m is related to the force **F** applied to this body, as

$$m\mathbf{a} = \mathbf{F}. \tag{3.1}$$

In Newtonian mechanics, force is a fundamental notion and it can not be defined precisely, nor it can be defined in terms of other notions. The vector of force, $\mathbf{F}$, applied to a particle, represents the effect of interaction with other particles and bodies. According to Eq. (3.1), this interaction manifests itself in the form of acceleration $\mathbf{a}$ acquired by the particle of a mass m.

It is assumed that all the forces in mechanics have the property called *law of superposition*. This means that if a particle is subject to several independent forces, e.g., $\mathbf{F}_1$, $\mathbf{F}_2$ and $\mathbf{F}_n$, its acceleration is the result of the action of the vector sum of these forces,

$$m\mathbf{a} = \mathbf{F}_1 + \mathbf{F}_2 + \ldots + \mathbf{F}_n. \tag{3.2}$$

The fact that the forces can be added and that their sum generates acceleration is not a trivial property. Indeed, for many forces in nature, this relationship is not valid. For example, among the four fundamental forces of nature (strong interactions, weak, electromagnetic and gravitational) only the electromagnetic interaction satisfies the property (3.2), while the weak interaction and especially the strong one violate this rule.

The gravitational interaction, in general, also violates the rule (3.2), but for weak gravitational fields the formula (3.2) can be used with great precision. The violent breakdown of the law of superposition, in the case of the gravitational field, is possible only for special objects like black holes, which can produce an extremely intense gravitational field. In the case of objects such as stars or galaxies, the effect may be detectable but is relatively weak. For this reason, many properties of astronomical objects can be studied within the scope of Classical Mechanics, where the rule (3.2) is regarded as a universal law.[1]

It proves useful to introduce the concept of resulting force, or resultant, $\mathbf{F}_R$,

$$\mathbf{F}_R = \mathbf{F}_1 + \mathbf{F}_2 + \ldots + \mathbf{F}_n, \tag{3.3}$$

and rewrite the Eq. (3.2) in the way $m\mathbf{a} = \mathbf{F}_R$. From now on, unless the opposite is explicitly stated, we assume that the force acting on a particle is a resultant force, that is, we call $\mathbf{F}_R$ simply $\mathbf{F}$. Equation (3.1) can be represented as a system of differential equations for the coordinates of the particle,

$$\ddot{x} = \frac{F_1}{m}, \qquad \ddot{y} = \frac{F_2}{m}, \qquad \ddot{z} = \frac{F_3}{m}, \tag{3.4}$$

where the force components depend, typically, on the particle's position and velocity, and on time,

$$F_i = F_i(x, y, z; \dot{x}, \dot{y}, \dot{z}; t), \qquad \text{where} \qquad i = 1, 2, 3, \tag{3.5}$$

but do not depend on the components of acceleration.

---

[1] One has to note, however, that the mentioned small corrections are in some cases relevant. For example, taking small corrections coming from General Relativity into account is a necessary condition for GPS to work with a proper precision.

Newton's Second Law, in conjunction with the third law, forms the basis for studying the dynamics of a single particle or a system of particles. Mathematically, the relations (3.4) are ordinary differential equations of the second order. The general solution of these equations always depend on a set of arbitrary constants of integration, which can be fixed by the use of initial conditions. In most cases, these conditions enable one to fix all the constants of integration and finally get a necessary particular solution. For example, the initial conditions may include the position and the velocity of a particle in a given initial instant. Typically, the initial conditions allow us to define the motion of a particle in a unique way. But there may be exceptions such as, for example, a particle placed on the saddle point of a curved surface, as it is shown in Fig. 3.1. If at the initial moment the speed of the particle is zero, then it can move either to the right or to the left, or stay in rest. In this case the initial conditions do not define a unique solution.

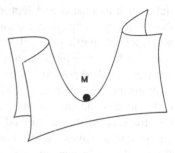

Fig. 3.1 Still particle in a saddle point

The movement of a particle, in the general case, is described by a system of three differential equations (3.5). In some cases it is useful to reduce the number of equations using the co-moving basis (see Chap. 2). With the basis vectors $\hat{\mathbf{n}}$ and $\hat{\mathbf{v}}$ attached to the particle, it is possible to write Newton's Second Law in the form

$$m\frac{dv_t}{dt} = F_t, \qquad m\frac{dv_n}{dt} = F_n,$$

where F_t and F_n are tangential and normal components of a force

$$\mathbf{F} = F_t\hat{\mathbf{v}} + F_n\hat{\mathbf{n}}.$$

The illustration can be seen in Fig. 3.2. An example of the use of this type of basis will be considered below.

3.4 Brief Classification of Forces in Mechanics

In this chapter we will provide the content to Newton's Laws, namely we briefly review the most important forces encountered in mechanics. We can always distinguish the microscopic forces that act between molecules, atoms, nuclei and elementary particles from the macroscopic forces, acting between macroscopic bodies.

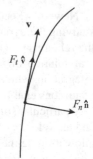

Fig. 3.2 Decomposition of
the force in the co-moving
basis

The microscopic forces can be viewed as being more fundamental, because they
act between smaller and more basic objects. They can be classified into fundamental and effective forces. The fundamental forces act between elementary particles such as quarks and leptons. The difference between quarks and leptons is
that quarks participate in all interactions (the strong, weak and electromagnetic)
while leptons participate only in the weak and electromagnetic (except neutrino)
ones. Each of these particles has its antiparticle. For example, the antiparticle of the
electron is called positron, it has electrical charge opposite to the electron charge.
Leptons include electrons and also similar particles (with the same electric charge)
called muons and tau-leptons, e, μ, τ. Furthermore, there are three neutrino types,
one for each of the mentioned leptons, ν_e, ν_μ, ν_τ. Neutrinos have no electrical
charge and are much lighter compared to e, μ and τ, but they are also leptons.
Being relativistic elementary particles, leptons and quarks may interact only via the
agents that transfer the force.[2] These agents can be light (even massless, because it
is not forbidden in relativistic theory) or heavy. When the mass of the agent is small
(or zero), the interaction is of a long range, whereas when its mass is larger, the
range of interaction is smaller. The gravitational and electromagnetic forces are mediated by massless particles (graviton and photon, correspondingly), so these forces
can be observed at macroscopic distances. The agents of the weak force are called
bosons W and Z and are very heavy. Therefore this force acts only at very small
distances. The elementary particles are relativistic and quantum objects, i.e., they
are not proper subjects for study within Classical Mechanics. In relativistic theory,
particles can be defined as massless like the photon – the quantum of the electromagnetic field. The elementary particles that participate in the strong interactions
are called quarks. There are six kinds of them: up, down, strange, charm, bottom
and top, normally indicated by u, d, s, c, b and t, respectively. The interaction between quarks is mediated by gluons. The strong interaction is called in this way
because it causes the phenomenon of confinement. This means that the quarks in
nucleons have such strong ties, that they can not be separated from each other. As a
result, we can not observe free quarks and can only obtain information about them

---

[2] See discussion of this issue in the Sect. 3.6 about Newton's Third Law.

in complicated experiments, e.g., on accelerators, where the involved energies are enormous. In ordinary matter, quarks form stable clusters, called nucleons (protons and neutrons), comprising the nuclei of atoms in addition to other particles called mesons. Regardless of the fact that both nucleons and mesons are composed of quarks, they play the role of fundamental elements at the nuclear physics scale. In fact, mesons transfer interactions between nucleons, similar to what W and Z bosons do at the higher, electroweak, energy scale. So, by observing these particles (outside of powerful accelerators) it is not easy to see that they are made out of quarks.

At scales of distances greater than the size of a proton, the strong interactions are no longer visible in contrast to the nuclear forces. Nuclear forces have their origin in the elementary strong interaction, but their properties are quite different, because quarks are hidden inside nucleons. The situation can be compared with the relations between two rival teams. Players from both teams can be friends, but it is hard to see this during the game. The same happens when we observe an interaction between two nuclei: this interaction may have very different properties compared to the interaction between quarks.

The nuclear forces are not fundamental, because they act between composite objects. For this reason they are considered to be *effective interactions*. When we increase the distance scale to the size of an atom, nuclear forces are also no longer observable and the only dominant force is the electromagnetic interaction. The traces of nuclear interactions can be seen only in the existence of different kinds of atoms. At the atomic or molecular scale, the forces, in general, do not obey the laws of Classical Mechanics, because quantum effects are very relevant. However, this does not imply that Classical Mechanics is completely useless in the field of microscopic physics. The application of Classical Mechanics for the description of movement of an electron, ion or an atom is indeed possible as an approximation, which can be valid (or not) for some particular problem. It is clear that the quality of such an approximation can be evaluated only by experiment or, theoretically, in the framework of a more general theory, e.g., Quantum Mechanics.

Anyway, the main field of application of Classical Mechanics is at the macroscopic scale. Below, we list the forces that often appear in problems of Classical Mechanics.

1) The gravitational force between two point masses, m_1 and m_2, separated by a distance r, is an attractive force, with magnitude given by $F_{12} = G\frac{m_1 m_2}{r^2}$, or, in vector form,

$$\mathbf{F}_{12} = -G\frac{m_1 m_2}{r^2}\hat{\mathbf{r}}, \quad \text{where} \quad \hat{\mathbf{r}} = \frac{\mathbf{r}_{12}}{r}, \quad r = |\mathbf{r}_{12}|. \tag{3.6}$$

The vector $\mathbf{r}_{12}$ is defined in Fig. 3.3. Henceforth we denote the force that the particle of mass m_1 exerts on the particle of mass m_2 as $\mathbf{F}_{12}$ and the force that the particle of mass m_2 exerts on the particle of mass m_1 as $\mathbf{F}_{21}$. According to Newton's Third Law (see Sect. 3.6), these two forces always satisfy the relation $\mathbf{F}_{12} = -\mathbf{F}_{21}$. One can see Fig. 3.3 for an illustration.

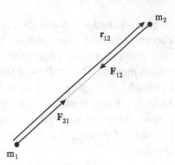

Fig. 3.3 Gravitational forces
between two particles

According to the formula (3.6) and Newton's Second Law, the acceleration of
the particle of mass m_1 does not depend on the value of this mass and is defined as

$$\mathbf{a}_1 = \frac{\mathbf{F}_{21}}{m_1} = \frac{G m_2 \hat{\mathbf{r}}}{r^2}. \tag{3.7}$$

It is important to mention that the masses of the objects that generate a significant
gravitational field (e.g., masses of planets, stars, galaxies, etc.) are not point-like, but
are distributed in space in some complicated way. In many cases these distributions
have approximate spherical symmetry. Therefore, in order to deal with the gravitational fields of these objects we need an evaluation of the gravitational field
generated by a spherically symmetric mass distribution. As we will show in the
Appendix of this chapter, a homogeneous spherical thin shell of mass M and radius
R generates a zero gravitational field in the region $r < R$ and a field with inten-
sity $|\mathbf{g}| = GM/r^2$ in the region $r > R$, where r is the distance to the center of the
shell. As a consequence, any body with spherical symmetry generates, outside its
own volume, the same gravitational field as the one generated by a point-like mass
placed in the center of the body.

If we consider the gravitational field,

$$\mathbf{g} = -\frac{GM}{r^2} \hat{\mathbf{r}},$$

in a region near the surface of a spherical planet, we can write the distance to the
center of the planet as $r = R + h$, where h is the height of the point with respect
to the surface of the planet with radius R. Considering $h \ll R$, we can expand the
factor $(R + h)^{-2}$ into Taylor series,

$$\frac{1}{(R+h)^2} = \frac{1}{R^2} - \frac{2h}{R^3} + \mathcal{O}\left(h^2\right). \tag{3.8}$$

To better understand the terms in the equation above, let us consider the gravitational
field near the surface of our planet. We remember that the Earth's radius is approx-
imately $R_T = 6{,}400\,\text{km} = 6.4 \cdot 10^6\,\text{m}$. For example, for the altitude $h = 10\,\text{km} = 10^4\,\text{m}$, the second term in the right side of Eq. (3.8) is 300 times smaller than the

first one, and the next terms of expansion are about 10^5 times smaller than the first one. Thus, for such "small" altitudes ($h \ll R$), we can neglect the second term and find the gravitational field to be approximately homogeneous

$$\mathbf{g} \cong -\frac{GM}{R^2}\hat{\mathbf{r}} = \text{const.}$$

near the Earth's surface.

2) The Coulomb force of attraction between two point charges is very similar to Newton's law for the gravitational attraction. Only the sign may be different depending on the signs of the electric charges,

$$\mathbf{F}_{12} = k\frac{q_1 q_2}{r^2}\cdot\hat{\mathbf{r}},$$

where $\hat{\mathbf{r}}$ is the same unit vector that appears in Eq. (3.6), illustrated in Fig. 3.3. If the charges have the same sign, there will be repulsion, if the signs are different, there will be attraction.

3) The magnetic force depends on the magnetic field, $\mathbf{B}$, and acts on a charged particle moving with the velocity[3] $\mathbf{v}$,

$$\mathbf{F} = q\,[\mathbf{v},\mathbf{B}]\,, \tag{3.9}$$

where q is the charge of the particle. According to the properties of the cross product, this force is always perpendicular to the velocity and is also perpendicular to the magnetic field.

If a particle of mass m is subject to electric and magnetic forces at the same time, the dynamical equation will be as follows:

$$m\mathbf{a} = q\,(\mathbf{E} + \mathbf{v}\times\mathbf{B})\,. \tag{3.10}$$

The right side of this equation is called Lorentz's force.

4) The force of elasticity has its origin in the electromagnetic interactions inside a deformed solid body. For small deformations and in relatively simple situations, this force is proportional to the deformation (see Fig. 3.4):

$$\mathbf{F} = -k\Delta\mathbf{r}. \tag{3.11}$$

This relation is called Hooke's Law. The constant k is the coefficient of elasticity (elastic constant). In the general case of deformed solid body the force caused by the deformation may have a different direction than the displacement of the corresponding point of the body. The deformations in these cases are dealt with in the part of Physics called Elasticity Theory. In our course, we will always assume the simplest law (3.11).

---

[3] Equation (3.9) is in MKS units. In the Gaussian system of units one has to replace $q \to q/c$, where c is the speed of light.

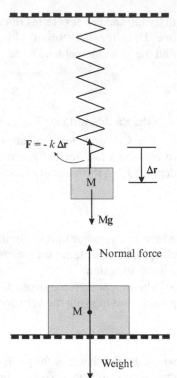

Fig. 3.4 Small distortions of a spring producing elastic force given by Eq. (3.11)

Fig. 3.5 A block on the floor. The normal force, together with the force of gravity, produces zero resulting force. The deformation of the base surface can be disregarded

In Classical Mechanics, in many situations, we consider the coefficient of elasticity to be infinite. This means that the strength force is finite while the deformation of the body is negligible. For example, when a body is placed on a floor surface, we can take the approach that the background surface is not deformed. The force that the ground exerts on the body is perpendicular to the surface, so it is called a normal force. The diagram of forces in this case is shown in Fig. 3.5. As the acceleration of the body is zero, in this particular case the sum of the forces acting on the body (weight and the normal force) is also zero. Consequently, the normal force has the same magnitude and is parallel to the gravitational (weight) force, but it has an opposite direction.

Another similar example is an idealization called an inextensible "wire". In this case, we assume that the stressed wire develops the necessary elasticity force without essentially changing its length. This is equivalent to establish an infinite coefficient of elasticity.

5) The frictional force between the two surfaces in relative motion arises from the mutual interaction between them. It is assumed to be proportional to the normal force between the surfaces and so its magnitude is given by $F_{atr} = kN$. The vector of the force always acts against the direction of the motion of the body, so it can be written as (see Fig. 3.6)

$$\mathbf{F}_{fri} = -kN\hat{\mathbf{v}}. \tag{3.12}$$

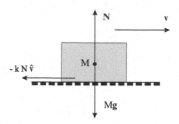

Fig. 3.6 A sliding block on the ground undergoes the action of force of friction, which resists the movement

6) The resistance of a fluid (including gas, e.g., air) is another kind of the frictional force. For small velocities it is proportional to the magnitude of the speed of the body, then

$$\mathbf{F}_{res} = -\alpha\mathbf{v}. \tag{3.13}$$

For higher speeds, there is a quadratic term in velocity. In principle, there may be also cubic terms and so on. We can write the general expression for the friction force as

$$\mathbf{F}_{res} = -k(v)\,\hat{\mathbf{v}}, \tag{3.14}$$

where $v = |\mathbf{v}|$ is the magnitude of the velocity of a particle. In the simplest case, considering (3.13), obviously, $k(v) = \alpha v$. Interestingly, the dynamic frictional force between two surfaces (3.12) also fits the general rule (3.14). The formula (3.14) can be viewed as a more general form of friction force, also called a dissipative force. Of course, the function $k(v)$ can be expanded in power series in velocity. In the case of resistance of air or another fluid in the small speed regime, the linear term (3.13) is the most important one. In this case the equations of motion are typically simpler. For higher speeds, the quadratic term begins to dominate and terms of higher orders may also be relevant. It is worth mentioning that if the fluid is an ideal gas, the resistance is due to elastic scattering of gas particles. In this case, the quadratic term is more important than the next expansion orders in the system for high speeds (see one of the next Exercises).

3.5 Applications of Newton's Second Law

Consider some simple examples of application of Newton's second law in Mechanics.

Example 1. A projectile, after a shot from a cannon, is launched with initial velocity v_0 and initial angle φ relative to the horizon. Find the equation of motion and the trajectory of the projectile ignoring the air resistance (see Fig. 3.7).

Solution. The equation of motion of the projectile, without taking into account the air resistance is given by

$$\frac{md\mathbf{v}}{dt} = m\mathbf{g} = -mg\,\hat{\mathbf{k}}.$$

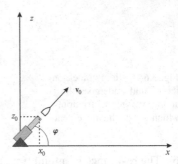

Fig. 3.7 Oblique launching of
a projectile

Making the decomposition $\mathbf{v} = v_x\hat{\mathbf{i}} + v_z\hat{\mathbf{k}}$, we obtain the two equations,

$$m\dot{v}_x = 0 \quad \text{and} \quad m\dot{v}_z = -mg,$$

with initial conditions $v_x(0) = v_0\cos\varphi$, $v_z(0) = v_0\sin\varphi$. Furthermore, the initial data for the own coordinates x and z are $x(0) = x_0$ and $z(0) = z_0$. Here, z_0 is the height of the cannon. Integrating over time, we find

$$v_x(t) \equiv v_0\cos\varphi, \quad v_z(t) = v_0\sin\varphi - gt$$

and finally

$$x = x_0 + v_0 t\cos\varphi, \qquad z = z_0 + v_0 t\sin\varphi - \frac{gt^2}{2}. \tag{3.15}$$

The last formulas represent the equations of motion. A special case is when the nozzle is held straight up, i.e., $\varphi = \pi/2$. We have

$$v_x \equiv 0, \quad v_z = v_0 - gt, \quad x \equiv x_0, \quad z = z_0 + v_0 t - \frac{gt^2}{2}.$$

To find the equation of trajectory for $\varphi \neq \pi/2$, one has to eliminate t in the Eqs. (3.15):

$$t = \frac{x - x_0}{v_0\cos\varphi},$$

$$z - z_0 = (x - x_0)\tan\varphi - \frac{g}{2v_0^2\cos^2\varphi}(x - x_0)^2. \tag{3.16}$$

We note that the trajectory is a parabola, contained in the same plane as the velocity vector and acceleration $\mathbf{g}$. The equation of motion and the trajectory do not depend on the mass of the particle. The last property is a direct consequence of the exact equality between the inertial and gravitational masses.

This example is very simple but at the same time it allows us to observe several properties of motion under the influence of the gravitational force. Furthermore, the example is useful for comparison with the more complicated case, when the resistance of air is taken into account.

Example 2. A projectile of mass m is launched vertically upwards. The force of air resistance is given by $F_r = -kv$, where $k = \alpha m$ is a constant coefficient.[4] Consider the motion of the projectile until the moment when it stops at the top. What would change if the projectile had been thrown down?

Solution. The problem is essentially one-dimensional because both the initial velocity and acceleration are parallel to the axis OZ,

$$\mathbf{v_0} = \pm v_0 \mathbf{\hat{k}}, \qquad \mathbf{a} = \mathbf{g} = -g\mathbf{\hat{k}}.$$

The positive sign corresponds to the upward movement and the negative one to the downward movement. We consider here only the positive sign and leave the other case as an exercise. The dynamical equation is

$$m\frac{d\mathbf{v}}{dt} = m\mathbf{g} - k\mathbf{v} = m\mathbf{g} - \alpha m\mathbf{v}.$$

Projecting the last equation on the axis OZ (which may be done through the scalar product with $\mathbf{\hat{k}}$), we obtain

$$m\frac{dv}{dt} = -gm - \alpha mv = -\alpha m\left(v + \frac{g}{\alpha}\right).$$

Let us note that when the movement is upward, both forces (gravitational and air resistance) have directions opposite to the velocity. The last relation is an ordinary differential equation of the first order. The equation is separable, i.e., we can write

$$\frac{dv}{v + g/\alpha} = -\alpha dt.$$

Integrating both sides in v and t correspondingly, we obtain

$$\ln\left|\frac{v + g/\alpha}{C}\right| = -\alpha t,$$

where C is an integration constant. Using the initial condition $v(0) = v_0$, we find $C = v_0 + g/\alpha$ and finally obtain the particular solution in the form

$$\ln\left|\frac{v + g/\alpha}{v_0 + g/\alpha}\right| = -\alpha t.$$

Solving the last relation for the velocity, we arrive at

$$v = -\frac{g}{\alpha} + \left(v_0 + \frac{g}{\alpha}\right)e^{-\alpha t} = v_0 e^{-\alpha t} + \frac{g}{\alpha}\left(e^{-\alpha t} - 1\right). \qquad (3.17)$$

---

[4] We use this representation for the friction coefficient k only to simplify the notations. In fact, k does not depend on mass, but only on the shape of a body and on the quality of its surface.

This equation carries important information about the motion of the projectile. It may be noted that the projectile moves up until the instant t_c, defined by $v(t_c) = 0$. To determine this value, one can write

$$e^{-\alpha t_c} = \frac{g}{\alpha v_0 + g}, \qquad \text{or} \qquad t_c = \frac{1}{\alpha} \ln \left| 1 + \frac{\alpha v_0}{g} \right|. \tag{3.18}$$

The equation of motion can be easily found by integrating the Eq. (3.17), valid on the time range $0 < t < t_c$:

$$z = z_0 - \frac{gt}{\alpha} + \frac{1}{\alpha} \left(v_0 + \frac{g}{\alpha} \right) \left(1 - e^{-\alpha t} \right). \tag{3.19}$$

The maximal height is

$$h = z(t_c) - z_0 = \int_0^{t_c} v(t)dt = \frac{v_0}{\alpha} - \frac{g}{\alpha^2} \ln \left| 1 + \frac{\alpha v_0}{g} \right|. \tag{3.20}$$

An interesting observation is the following. We can expect that when the air resistance disappears, the formulas above should reduce to the same as discussed in the first Example, in the case where $\varphi = \pi/2$. At the same time the direct inspection allows us to conclude that all expressions (3.17)–(3.20) are singular (this means they tend to infinite values) when we consider the limit $\alpha \to 0$. Let us show the solution of this problem for the velocity (3.17). For small values of α one can expand the exponential in power series in α, according to

$$e^{-\alpha t} = 1 - \alpha t + \frac{(\alpha t)^2}{2} - \frac{(\alpha t)^3}{3!} + \cdots.$$

Now, replacing this expansion into (3.17), we observe that the negative powers of α disappear. Keeping the terms up to the first order in α, we obtain

$$v(t) = v_0 - gt - \alpha \left(v_0 t - \frac{gt^2}{2} \right) + \mathcal{O}(\alpha^2). \tag{3.21}$$

It is easy to see that the speed of the projectile is always smaller than the one calculated without air resistance. We leave the analysis of the same problem for $z(t)$ and the maximum height of ascent of the projectile h, as an exercise.

Example 3. A projectile of mass m is launched with initial velocity v_0 and initial angle φ relative to the horizon. The force of air resistance is the same as in the previous problem, $\mathbf{F}_r = -\alpha m \mathbf{v}$. Determine the motion of the projectile.

Solution. This problem can be seen as a combination of the two previous problems. We will follow the same scheme of solution. The equation of motion and the initial conditions are given by

$$m\frac{d\mathbf{v}}{dt} = m\mathbf{g} + \mathbf{F}_r = m\mathbf{g} - k\mathbf{v} = -mg\hat{\mathbf{k}} - m\alpha\mathbf{v},$$

$$v_x(0) = v_0\cos\varphi, \quad v_z(0) = v_0\sin\varphi,$$

$$x(0) = x_0, \quad z(0) = z_0. \tag{3.22}$$

Projecting the equation of motion onto the axes OX and OZ, we obtain

$$\frac{dv_x}{dt} = -\alpha v_x, \qquad \frac{dv_z}{dt} = -g - \alpha v_z. \tag{3.23}$$

Now we can take advantage of our system of linear equations, which are decoupled, or, in mathematical language, factorized. It may be noted that the equation for $v_z(t)$ is the same as appears in Example 2, with obvious modifications from the initial condition, $v_0 \to v_0\sin\varphi$. Then, the solution of this equation can be immediately obtained from the Eq. (3.17) in the form

$$v_z(t) = -\frac{g}{\alpha} + \left(v_0\sin\varphi + \frac{g}{\alpha}\right)e^{-\alpha t}. \tag{3.24}$$

The solution of the equation for $v_x(t)$ has been discussed in Chap. 2, with the result given by

$$v_x(t) = v_0\,e^{-\alpha t}\cos\varphi. \tag{3.25}$$

The formulas (3.24) and (3.25) admit an integration and enable one to find the functions $x(t)$ and $z(t)$ and the equation of the trajectory $z(x)$, that we leave as an exercise.

Example 4. A small body of mass m is sliding on a sphere of radius R, frictionless (Fig. 3.8). Explore the movement of the body and find its speed at the time when it loses contact with the surface of the sphere.[5]

Fig. 3.8 Illustration of Example 4. The solution to this problem is simpler using energy conservation (it will be discussed later), but here we consider the solution directly by Newton's Second Law

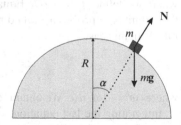

Two forces act on the body: the gravitational force, $m\mathbf{g}$, and the normal force, $\mathbf{N}$. One can write Newton's Second Law in the co-moving base, $(\hat{\mathbf{n}}, \hat{\mathbf{v}})$, remembering that the formula for the centripetal acceleration is given by $a_n = v^2/R$.

---

[5] This problem is from Irodov's book [16], but here it has been solved with more detail.

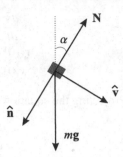

Fig. 3.9 Diagram of forces
and the co-moving base
(Example 4)

According to the Fig. 3.9 we have

$$ma_n = mg\cos\alpha - N \quad \text{and} \quad ma_t = m\frac{dv}{dt} = mg\sin\alpha.$$

To solve the problem, remember that $v = R\omega = R\dot{\alpha}$. Then, the second equation reduces to a second-order differential equation for the angle of deflection from equilibrium,

$$\ddot{\alpha} = \frac{g}{R}\sin\alpha,$$

and the first equation defines the normal force,

$$N = m(g\cos\alpha - R\dot{\alpha}^2).$$

The initial conditions of the differential equation for the angular variable is given by $\alpha(0) = 0$ and $\dot{\alpha}(0) = 0$.

Then we obtain

$$\ddot{\alpha} = \frac{g}{R}\sin\alpha = f^2\sin\alpha, \tag{3.26}$$

where we define $f = \sqrt{g/R}$. Equation (3.26) is a second-order differential equation that admits the procedure called reduction of the order. Let us consider $\omega = \dot{\alpha} = \omega(\alpha)$. In this case

$$\ddot{\alpha} = \frac{d\omega}{dt} = \frac{d\omega}{d\alpha}\cdot\frac{d\alpha}{dt} = \omega\omega',$$

where $\omega' = d\omega/d\alpha$. We obtain $\omega\omega' = f^2\sin\alpha$, or $d(\omega^2) = 2f^2\sin\alpha\,d\alpha$.

Integrating the last equation, we find $\omega^2 = C - 2f^2\cos\alpha$, where C is a constant of integration. Taking into account that $\omega(\alpha = 0) = 0$, we conclude that $C = 2f^2$ and thus $\omega^2 = 2f^2(1 - \cos\alpha) = 4f^2\sin^2\frac{\alpha}{2}$. Assuming $\omega > 0$, we obtain $\omega = 2f\sin\frac{\alpha}{2}$, that is,

$$\frac{d\alpha/2}{\sin\alpha/2} = f\,dt.$$

The integration of this equation is simple. We can perform the following manipulations:

$$\frac{d(\alpha/2)}{\sin \alpha/2} = -\frac{d \cos \alpha/2}{1 - \cos^2 \alpha/2} = \frac{1}{2} d \ln \left| \frac{\cos \alpha/2 - 1}{\cos \alpha/2 + 1} \right|.$$

Then,

$$\frac{1}{2} \ln \left| \frac{\cos \alpha/2 - 1}{\cos \alpha/2 + 1} \right| = f(t - t_0), \tag{3.27}$$

where t_0 is an integration constant. The analysis of this solution can be quite instructive. If one takes $\alpha = 0$, it is easy to see that this corresponds to an infinite initial time, $t_0 = \infty$. This means that if the body is in the position $\alpha = \omega = 0$ at the initial instant, it remains in this position forever. At the same time, a small deviation from the equilibrium position or small initial angular velocity ω already implies in a movement of the particle. The Eq. (3.27) can be solved with respect to α. The solution is given by

$$\alpha = 2 \arccos \left[\frac{e^{2f(t-t_0)} - 1}{e^{2f(t-t_0)} + 1} \right]. \tag{3.28}$$

In order to define the moment when the body will leave the surface, one can directly use the equations

$$N = mg \cos \alpha - \frac{mv^2}{R} = 0 \implies v^2 = gR \cos \alpha. \tag{3.29}$$

Using

$$\omega^2 = \frac{2g}{R} (1 - \cos \alpha) = \frac{v^2}{R^2},$$

we find

$$\cos \alpha = 1 - \frac{v^2}{2gR}. \tag{3.30}$$

Comparing (3.29) to (3.30), we find

$$\cos \alpha = 1 - \frac{gR \cos \alpha}{2gR} = 1 - \frac{\cos \alpha}{2},$$

that is, $\alpha = arc \cos \frac{2}{3}$. Also, using Eq. (3.30), we can conclude that $v = \sqrt{\frac{2gR}{3}}$ at the initial instant of the free fall of the body.

Exercises

1. Repeat the consideration of Example 2 for the movement pointed down, that is, for the case $v_0 = -v_0\hat{k}$. Explain using equations and also qualitatively, why in this case there is no maximum time t_c. Consider the asymptotic behavior $t \to \infty$. Explain why there is a speed limit in this case. How can one calculate its value without solving any differential equation?

Solution. The differential equation for $v(t)$, in this case, has the form

$$\frac{dv}{v - g/\alpha} = -\alpha dt, \quad v(0) = v_0. \tag{3.31}$$

The solution can be easily obtained in the form

$$v = \frac{g}{\alpha} + \left(v_0 - \frac{g}{\alpha}\right)e^{-\alpha t} \to \frac{g}{\alpha} \quad \text{for} \quad t \to \infty.$$

This asymptotic behavior can be seen directly by inspecting Eq. (3.31), because g/α is the only value of v, which makes the integral on the right side of Eq. (3.31) infinite. This type of solution is called "fixed point" of the equation. From the physical viewpoint, the value $v = g/\alpha$ corresponds to the equilibrium between gravity and air resistance forces and can be obtained by elementary means.

2. Perform the following considerations:

(a) Conclude the calculations of Example 3, i.e., find the functions $x(t)$ and $z(t)$, and the equation of the trajectory, $z(x)$.
(b) Consider the initial speed downwards, i.e.,

$$v_x(0) = v_0 \cos\varphi, \qquad v_z(0) = -v_0 \sin\varphi.$$

For this case, repeat the consideration and discuss its similarity to the result of Exercise 1.

Answer:

(a) $\quad x = x_0 + \dfrac{v_0 \cos\varphi}{\alpha}\left(1 - e^{\alpha t}\right), \quad z = z_0 - \dfrac{gt}{\alpha} + \dfrac{1}{\alpha}\left(\dfrac{g}{\alpha} + v_0 \cos\varphi\right)\left(1 - e^{\alpha t}\right).$

3. In Example 3, the body rises and then falls, until it returns to the initial height. Calculate the time intervals of ascent and descent movements. Which one is bigger and why? Try to get the answer without doing calculations, by means of qualitative considerations. Then try to check the result using equations.

Answer: For the rise time, t_{up}, we have

$$t_{up} = \frac{1}{\alpha} \ln\left(1 + \frac{\alpha v_0 \sin\varphi}{g}\right).$$

The fall time t_{down} satisfies the equation

$$\alpha t_{down} + e^{-\alpha t_{down}} = \frac{\alpha v_0 \sin \varphi}{g} + \ln\left(1 + \frac{\alpha v_0 \sin \varphi}{g}\right).$$

Considering the difference between these two time intervals, we obtain

$$t_{down} - t_{up} = \frac{1}{\alpha}\left[1 + x - 2\ln(1+x)\right] - \frac{1}{\alpha}e^{-\alpha t_{down}}, \quad \text{where} \quad x = \frac{\alpha v_0 \sin \varphi}{g}.$$

This equation can not be solved analytically in general form. Qualitatively, it is obvious that $t_{down} > t_{up}$, because the magnitude of the acceleration is greater when the projectile is moving upward, at equal heights.

3.6 Newton's Third Law

Newton's Third Law can be formulated as follows. Consider two interacting bodies, 1 and 2. If $\mathbf{F}_{12}$ is the force that the body 1 exerts on the body 2, and if $\mathbf{F}_{21}$ is the force that the body 2 exerts on the body 1, then these two forces will have the same magnitude and exactly opposite directions,

$$\mathbf{F}_{12} = -\mathbf{F}_{21}. \tag{3.32}$$

The situation is illustrated in Fig. 3.3.

The third law has several interesting aspects. Before studying the use of this law, it is worthwhile to discuss its relationship to the postulates of Mechanics. It may be noted that the third law requires instantaneous interaction and consequently the transfer of interaction with an infinite speed. In order to better understand this issue, imagine the opposite case, in which, for example, the gravitational interaction is transferred with finite speed, equal to c. Consider the interaction between the Earth and the Moon. The transfer time of a signal in this case is given by the distance between the Earth and the Moon (about 384,000 km) divided by the speed of 300,000 km/s, i.e., something close to 1.3 s. We assume, for the sake of simplicity, that the motion of the Earth is negligible and only the Moon is moving by the orbit around the Earth. We can calculate, easily, that during the time that the gravitational signal from the Earth reaches the Moon, the Moon travels a distance of about 1.19 m. So, if we assume a finite speed of the signal, we must admit that for the force exerted by the Moon on the Earth and the force exerted by Earth on the Moon, there will be always a small but nonzero angle, due to the displacement of the Moon.

Therefore, in the theory of relativity, with a maximum transfer speed for the signal, we can not accept the interaction by distance and there should be always a kind of mediator object transferring the force. For the electric and magnetic forces, corresponding to electromagnetic phenomena, such a mediator is the photon, that is the quantum of the electromagnetic field. In gravitation (General Relativity), the quantum of the field is called graviton. The process of a graviton exchange between two point particles is shown in Fig. 3.10 (this kind of figure is called Feynman diagram). In Quantum Field Theory, Newton's gravitational law (3.6) can be easily obtained by consideration of such an exchange process. For weak and strong interactions, mediators are massive and thus the interactions can only be observed at very small distances. When we accept that there are some special particles that mediate interactions, Newton's third law becomes perfectly compatible with Relativity.

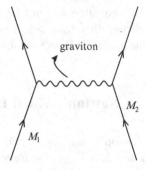

Fig. 3.10 Gravitational interaction between two particles mediated by graviton

Let us come back to Classical Mechanics, where there is no notion of a mediator of the interaction. Then the only way to solve the contradiction and ensure the validity of the Third Law is assigning an infinite speed for the propagating information. This is equivalent to using an approximation of very small speeds compared to the speed of light. In this context, it is possible to consider the interaction at a distance in the non-relativistic physics.

Let us now consider some simple examples of using Newton's Third Law. Although the level of these examples corresponds to high school, they are important to illustrate the meaning of the Third Law.

Example 5. Consider a horizontal plane with a non-deformable block of mass m on it. The forces acting on the block are the weight force $\mathbf{P} = m\mathbf{g}$ and the normal force $\mathbf{N}$, exerted by the plane. As the block is at rest, by Newton's Second Law the sum of these forces is zero, and so we obtain $\mathbf{N} = -m\mathbf{g}$. According to Newton's Third Law, the force that the block exerts on the plane is given by $-\mathbf{N} = m\mathbf{g}$, called the reaction of the normal force (which is called the action force). Note that forces $\mathbf{N}$ and $m\mathbf{g}$, although they have the same modulus and opposite directions, **do not** form a couple "action and reaction". The force of reaction to $m\mathbf{g}$ is the gravitational attraction exerted on the Earth by the block. What is the real reaction force to $\mathbf{N}$?

Example 6. The moon is rotating around the Earth on a circular orbit. The force that keeps the moon in this orbit is the centripetal force and, according to Newton's Second Law,

$$\mathbf{F}_c = M\mathbf{a}_c = -\frac{M v^2}{r}\hat{\mathbf{r}}, \qquad \hat{\mathbf{r}} = \frac{\mathbf{r}}{r}, \tag{3.33}$$

where **r** is the position vector of the Moon in the coordinate system with origin at the center of the Earth, v is the absolute value of the velocity of the Moon, $\mathbf{a}_c$ is the centripetal acceleration of the Moon and M is the mass of the Moon.

According to Newton's Third Law, the Earth is subject to the force $-\mathbf{F}_c$ exerted by the Moon, so the Earth also undergoes the acceleration

$$\mathbf{a}_c^E = -\frac{\mathbf{F}_c}{M_E} = -\frac{M}{M_E}\mathbf{a}_c. \tag{3.34}$$

We do not observe this acceleration only because the ratio of the masses of Moon M and of Earth M_E is very small. In the next chapter we will consider this situation in a general way, by discussing the so-called problem of two bodies.

Exercises

1. Consider the following problems with an inclined plane, without friction (see Fig. 3.11). In all cases, construct diagrams of forces and find the accelerations of all the blocks and tensile forces on all wires. All slopes have fixed positions.

Answers: We present the results for accelerations in three cases and the tension in one case,

$$(a) \quad a = g\sin\alpha;$$

$$(b) \quad a = \frac{m - M\sin\alpha}{m + M}g, \qquad T = \frac{gmM(1 + \sin\alpha)}{m + M};$$

$$(d) \quad a = \frac{m\sin\beta - M\sin\alpha}{m + M}g.$$

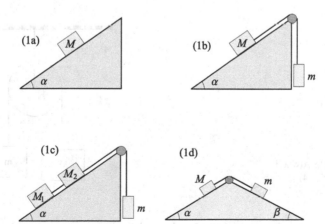

Fig. 3.11 Illustrations for the different cases of Exercise 1

2. In all cases from the previous exercise, find (if possible) the equilibrium conditions of the system. What are the possible motions under equilibrium conditions?

Answers: The equilibrium conditions of each system correspond to zero acceleration. If these conditions are met, the system would remain in equilibrium (at rest), but can also move with constant speed.

3. Generalize the results of previous exercises including a friction with a small coefficient μ between the blocks and surfaces. In the cases 1b, 1c and 1d (illustrated in Fig. 3.11), discuss the conditions of acceleration in both senses.

Hint. The equilibrium conditions in this case have the form of inequality.

Answers: We present the results only for cases *(a)* and *(d)*,

$$(a) \;\; a = g(\sin\alpha - \mu\cos\alpha);$$
$$\text{Equilibrium:} \quad a \le 0, \quad \text{i.e.,} \quad \mu \ge \tan\alpha.$$

$$(d) \;\; a = \frac{M\sin\alpha - m\sin\beta - \mu(M\cos\alpha + m\cos\beta)}{m+M}g$$
$$\text{in the case} \quad M\sin\alpha - m\sin\beta > 0.$$
$$\text{Equilibrium:} \quad a \le 0.$$

4. Consider a simple Atwood machine, frictionless (Fig. 3.12). Calculate the acceleration of the blocks and the tension force of the wire.

Answer:

$$a = \frac{m-M}{m+M}g, \quad T = \frac{2gmM}{m+M}.$$

(Compare with the result in Exercise 1b).

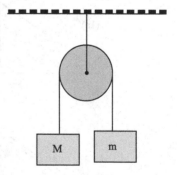

Fig. 3.12 Simple Atwood machine (Exercise 4)

5. Calculate the acceleration of the blocks and the tensions on the wires for the systems illustrated in Fig. 3.13.

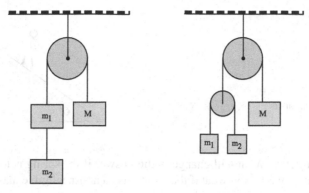

Fig. 3.13 Modified versions of the Atwood machine (Exercise 5). The diagram at the right side admits a simpler resolution in a non-inertial reference frame. We will consider this option in Sect. 3.7.1, where the answer can be found in Eqs. (3.44) and (3.45)

6. Calculate the tensions of the wires for the system illustrated in Fig. 3.14.

Answer:

$$T_\alpha = \frac{mg \cos\beta}{\sin(\alpha+\beta)}, \qquad T_\beta = \frac{mg \cos\alpha}{\sin(\alpha+\beta)}.$$

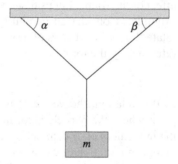

Fig. 3.14 Illustration of Exercise 6

7. A climber ascends the mountain, as shown in Fig. 3.15. The coefficient of friction between his shoes and the surface is $\mu = 0.2$. What is the maximum angle of inclination of the mountain for which he can go this way?

Hint. Use the result of the Exercise 3a.

8. Two blocks of masses m and M are connected by a spring of negligible mass and coefficient of elasticity k. The system is closed, i.e., there are no external forces. The spring is stretched and the system begins to oscillate without friction. Knowing that the maximal acceleration of the block of mass M is given by a_1, calculate maximal acceleration a_2 of the block of mass m. What will change in the answer

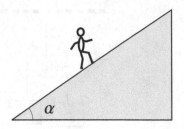

Fig. 3.15 Illustration for
Exercise 7

if the system rotates? What will change in the answer if the spring is too long, but
also of negligible mass? And what if the spring has a non-negligible mass?

Answers: The relationship $\mathbf{a}_2 = -(M/m)\mathbf{a}_1$ is valid for all situations where the
mass of the spring can be considered negligible and it is possible to assume that the
force is transferred by the spring instantaneously. It is clear that the second condition
is the consequence of the first one.

3.7 Non-inertial Reference Frames and Inertia Forces

Sometimes it is more useful to study the motion of a particle in a non-inertial reference frame. In this case, Newton's Second Law must be modified by including the
effect of the frame. Let us first consider the general situation and then some specific
examples. Consider an inertial system K and the non-inertial one K'. One can
relate the radius-vectors of a particle of mass m with respect to the two frames of
reference by the formula

$$\mathbf{r}' = \mathbf{r} - \mathbf{R}. \tag{3.35}$$

In this relation, the vector $\mathbf{R}$ is the difference between the radius-vectors of the
point where the particle is located, in the two frames. In the particular case when
the movement of system K' with respect to system K is purely translational (i.e.,
there is no rotation of the axes of K'), the vector $\mathbf{R}$ will be the same for all points in
space, and it can be written as $\mathbf{R} = \mathbf{OO}'$ (see Fig. 3.16), in this case, $\mathbf{R}$ is the vector
that joins the origins of the two coordinate systems. For more complicated relative
motions of K', the vector $\mathbf{R}$ can depend on the space point.

To determine the relationship between the accelerations of the particle in both
reference frames, we consider, at the instant t, the acceleration of system K' with
respect to reference system K. Considering the particle of the mass m, we denote
by $\mathbf{a}'$ the acceleration of this particle relative to K', and we introduce a new notation,

$$\mathbf{a}_p = \ddot{\mathbf{R}}, \tag{3.36}$$

where the vector $\mathbf{R}$ was defined in (3.35). Note that the acceleration $\mathbf{a}$, with respect
to K, can be different from $\mathbf{a}_p = \ddot{\mathbf{R}}$, due to the particle's acceleration in the non-
inertial reference frame, $\mathbf{a}'$.

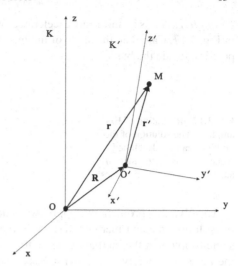

Fig. 3.16 Position of a particle in two reference frames in relative translational motion

The relation between the accelerations can be obtained by taking a second derivative of Eq. (3.35),

$$\mathbf{a}' = \mathbf{a} - \mathbf{a}_p. \tag{3.37}$$

Thus, Newton's Second Law, $m\mathbf{a} = \mathbf{F}$, can be written in the non-inertial frame K', in a modified form

$$m\mathbf{a}' = m\mathbf{a} - m\mathbf{a}_p = \mathbf{F} + \mathbf{F}_{in}. \tag{3.38}$$

The force $\mathbf{F}_{in} = -m\mathbf{a}_p$ is called force of inertia, or inertial force. It is introduced to compensate the acceleration of the system K' relative to K.

3.7.1 Accelerated Translational Motion

Let us now consider the inertial force in two particular cases of great practical importance. The first example is the case in which K' is experiencing a uniform acceleration $\mathbf{a}_p$ without changing the orientation of the axes with respect to K i.e., without rotation of the axes. An example of this situation is the acceleration of a car or bus in a straight path. When a person is inside such a bus, he/she feels the force of inertia in the opposite direction to $\mathbf{a}_p$. The same is true for any object or device inside the vehicle. Therefore, we can write

$$m\mathbf{a}' = \mathbf{F} - \mathbf{F}_{in} = \mathbf{F} - m\mathbf{a}_p. \tag{3.39}$$

At this point, let us remember the principle of equivalence of inertial and gravitational forces, as qualitatively discussed in the section dedicated to Newton's First Law. As we have already mentioned, this principle is very useful for solving certain problems of mechanics. Let us consider some examples.

Example 7. A bus is uniformly accelerated, with acceleration given by $\mathbf{a}_p$, as shown in Fig. 3.17. Calculate the angle of inclination, φ, of a pendulum at its equilibrium position inside the bus.

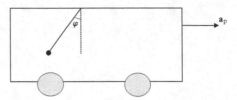

Fig. 3.17 Illustration of Example 7. The solution of the problem can be developed independently in the inertial and in the non-inertial frames

To solve this problem, we apply two independent approaches. First, we discuss the solution in a non-inertial reference frame, attached to the bus, and then solve the same problem in the inertial frame. In the first case, there are three forces acting on the mass m, namely   (1) gravity, $\mathbf{F}_{grav} = m\mathbf{g}$;   (2) force of inertia, $\mathbf{F}_{in} = -m\mathbf{a}_p$; (3) tension of the thread, $\mathbf{T}$. The sum of these forces is zero, because the mass is at rest relative to the bus. The diagram of forces is shown in Fig. 3.18.

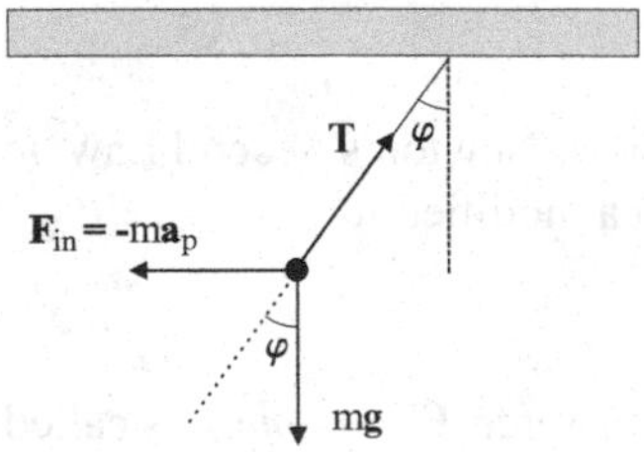

Fig. 3.18 Diagram of forces in non-inertial reference frame

Projecting the forces into horizontal and vertical axes, we obtain

$$T \sin \varphi = ma_p, \qquad T \cos \varphi = mg.$$

Thus, the tangent of the requested angle is given by $\tan \varphi = a_p/g$.

To solve the problem in the inertial frame, we remember that, in this case, there is no force of inertia. Therefore, one only need to sum up $\mathbf{T}$ with $m\mathbf{g}$ and write Newton's Second Law, remembering that the pendulum is accelerated together with the bus,

$$m\mathbf{g} + \mathbf{T} = m\mathbf{a}_p.$$

Of course, this is mathematically equivalent to the equation

$$m\mathbf{g} + \mathbf{T} + \mathbf{F}_{in} = 0,$$

found in the non-inertial reference frame. The result for $\tan \varphi$ is obviously the same.

Example 8. A bucket of water has, inside it, a wooden body of volume V, which is completely immersed in the water. The density of the body, ρ_b, is smaller than the density of water, ρ_w. However, the body does not emerge due to a thread which is connected to the bottom of the bucket. Calculate the tension on the wire in the two following situations: (a) The bucket is an inertial frame. (b) The bucket is an elevator that is rising up with constant acceleration a.

Solution. In the case (a) the body is a subject to three forces, namely gravity, $mg = V\rho_b g$, acting downwards, thrust force, $V\rho_w g$, acting upwards, and the tension in the thread, T_1, acting downwards. The sum of these forces is zero because the body is in rest. Then, Newton's Second Law takes the form

$$V\rho_b g + T_1 = V\rho_w g, \quad \text{so that} \quad T_1 = Vg(\rho_w - \rho_b). \qquad (3.40)$$

The case (b) is much more interesting. First of all, the law of Archimedes can not be applied to dynamical systems, such as the one moving with acceleration, because it is a law of hydrostatics. So the most convenient way to solve the problem is to use the non-inertial reference frame, where the bucket is at rest. In this case, we can use the equivalence between gravitation and inertia, reaching the conclusion that the *effective* gravitational acceleration inside the elevator is given by $\mathbf{g}_{ef} = \mathbf{g} - \mathbf{a}$, whose absolute value is $g_{ef} = g + a$. All processes inside the elevator are sensitive to the effective gravitational acceleration g_{ef}, and not to the conventional value, g. Therefore, the result for the thread tension can be obtained by Eq. (3.40), with the replacement $g \to g_{ef} = g + a$. In this way we arrive at the result

$$T_2 = V(g + a)(\rho_w - \rho_b). \qquad (3.41)$$

Exercise 1. Solve the same problem in the inertial (rest) frame. Check that the result is the same, given by the formula (3.41). Note that the previous solution (in the non-inertial reference frame) is much simpler.

Example 9. We use the non-inertial reference frame to solve the problem of the double Atwood machine, which was the second part of Exercise 5 in Sect. 3.5. The traditional solution to this problem can be found, for example, in the book [15], but it is much more troublesome.

Solution. Denote the tension of the thread, attached to the block of mass M as $\mathbf{T}$ and the tension of the thread attached to the blocks of masses m_1 and m_2 as $\mathbf{T}_1$. The projections of all vectors are considered positive if the vectors are directed upwards. Newton's Second Law applied to the block of mass M provides

$$-Mg + T = Ma. \qquad (3.42)$$

This means that the mobile pulley has acceleration a downwards, i.e., this pulley and the two blocks of masses m_1 and m_2 are subjects to the "effective" gravitational field given by $g_{ef} = g - a$. Therefore, Newton's equations for these two blocks have the same forms as the equations for the simple Atwood machine, but with g_{ef} instead of g. In this way we obtain

$$m_1(g-a)-T_1=m_1\tilde{a}_1, \quad m_2(g-a)-T_1=-m_2\tilde{a}_1, \qquad (3.43)$$

where $\tilde{a}_1=a_1+a$ is the acceleration of the block m_1 related to the mobile pulley. The Eqs. (3.43) have the same solution as the simple Atwood's machine, namely

$$\tilde{a}_1=\frac{m_1-m_2}{m_1+m_2}(g-a), \quad T_1=\frac{2m_1m_2(g-a)}{m_1+m_2}. \qquad (3.44)$$

The tension of the threads does not change when we change the frame, therefore the same value T_1 can be used in the Newton's equation for the mobile pulley. As this pulley has mass zero, the sum of the forces applied to it has to be zero, hence $T=2T_1$. Now, we can use Eq. (3.42) to find

$$a=\frac{4m_1m_2-M(m_1+m_2)}{4m_1m_2+M(m_1+m_2)}g. \qquad (3.45)$$

The acceleration of blocks of masses m_1 and m_2 can be calculated using $\tilde{a}_1=a_1+a$, together with (3.44) and (3.45).

Exercise 9 (Infinite Atwood Machine)

In the system shown in Fig. 3.19, there is a block of mass m on the wire on the left and on the other side an ideal pulley with a body of the same mass on the left side and on the other side the same arrangement, *ad infinitum*. Calculate the acceleration of the first block on the left side, specifying whether it is directed up or down.

Answer: $a=g/2$.

Hint.   Use the fact that the situation in the second pulley is the same as in the first, but changing $g\to g_{ef}=g-a$. The solution becomes very simple. Let us note that the solution of this problem in the inertial reference frame is very complicated.

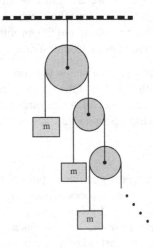

Fig. 3.19 Infinite Atwood machine (Exercise 9)

3.7.2 Uniform Rotational Motion

The second case, also of particular importance, is about a frame, K', which is rotating with constant angular speed ω around a fixed axis relative to an inertial frame, K. As we already know from the previous chapter, if we put the origins of the two frames at the same point on the axis of rotation, we will have the following relation between the accelerations of a point with the radius-vector $\mathbf{r}$ (remember that in this case $\mathbf{r}' = \mathbf{r}$):

$$\mathbf{a}' = \mathbf{a} + 2[\mathbf{v}', \vec{\omega}] - [\vec{\omega}, [\vec{\omega}, \mathbf{r}]].$$

Correspondingly, Newton's second law in the frame K' is modified by the inclusion of the inertial force,

$$ma' = ma + 2m[\mathbf{v}', \vec{\omega}] - m[\vec{\omega}, [\vec{\omega}, \mathbf{r}]]$$
$$= ma + 2m\mathbf{v}' \times \vec{\omega} - m\vec{\omega}\,(\vec{\omega}, \mathbf{r}) + mr\omega^2. \tag{3.46}$$

The last formula has four terms, each one with its own physical interpretation. For example, the first term, ma, is the only one that does not depend on the choice of the reference frame, so it is the only one that does not represent a force of inertia.

To better understand the interpretation of the terms on the right side of the Eq. (3.46), we consider first a special case where $\mathbf{r} \perp \vec{\omega}$. This means that the movement is flat (two-dimensional). In this case, the third term is zero, and we can write

$$ma' = ma + 2m\mathbf{v}' \times \vec{\omega} + m\omega^2 \mathbf{r}. \tag{3.47}$$

The last term is called the centrifugal force. It is well known by everybody who have been in curves inside a car or a bus. This term has the same direction as the position vector, $\mathbf{r}$. The term $m\vec{\omega}\,(\vec{\omega}, \mathbf{r})$ in the Eq. (3.46) does not admit any independent interpretation. The role of this term is to ensure that its sum with the centrifugal force (in the general case, when the condition $\mathbf{r} \perp \vec{\omega}$ is not satisfied) is orthogonal to the angular velocity vector, in accordance with the first form of the Eq. (3.46). Finally, the term

$$F_{Cor} = 2m\mathbf{v}' \times \vec{\omega} \tag{3.48}$$

is called the Coriolis inertial force. It is responsible for the asymmetry of the appearance of the rivers on the surface of the Earth and the difference in this respect between the northern and southern hemisphere. The physical interpretation of this force can be seen in the following manner. Let us consider the simplified case in which $\mathbf{v}' \perp \vec{\omega}$ and also $\mathbf{r} \perp \vec{\omega}$, i.e., the motion of the particle occurs in the plane perpendicular to the axis of rotation. If the velocity of a particle relative to K', $\mathbf{v}'$, has the same direction as the rotational speed of the same reference frame, i.e., $\mathbf{v}_{rot} = \mathbf{r} \times \vec{\omega}$, the effect of this speed will simply increase or decrease the centrifugal force, depending on the direction of $\mathbf{v}'$. In another case, when $\mathbf{v}'$ has the same direction as the radius-vector, $\mathbf{v}'\|\mathbf{r}$, the displacement of the particle along a radial segment will mean its speed will increase. For example, during the time interval Δt, the distance between the particle and the axis of rotation is increased by a value $\Delta r = v'\Delta t$. As a result, the rotation speed will increase from ωr to $\omega(r+\Delta r)$. The Coriolis force is the origin of this acceleration.

Example 10. A train of the mass m is moving along a circular trajectory, whose center coincides with the center of the Earth, at the latitude ξ, with constant velocity, v'. According to Fig. 3.20, the vector $\mathbf{v}'$ makes an angle α with the meridian. Determine the lateral force acting on the rails.

Solution. We need to determine the expression for the quantity $m\mathbf{a}$ in the Eq. (3.46), which represents the force exerted on the train. The component of the force which is tangential to the Earth's surface can only be exercised by the rails, because there is no such component in the weight force. In this way, we define $\mathbf{N} = \mathbf{N}_t + N_r\hat{\mathbf{n}}_r = -m\mathbf{a}$, where $\mathbf{N}_t$ is the tangential (to the surface of the planet, *not* only to the direction of motion) component of the force. It is necessary to calculate the vector component of $\mathbf{N}_t$ which is perpendicular to the rails, $\mathbf{N}_\perp$, i.e., perpendicular to the vector $\mathbf{v}'$.

To solve the problem, one should use the orthogonal basis in spherical coordinates (see Fig. 2.8). Note that the angular coordinate θ is given by $\theta = 90° - \xi$. We have, then

$$\mathbf{N} = 2m\mathbf{v}' \times \vec{\omega} - m\vec{\omega}(\vec{\omega}\cdot\mathbf{r}) + m\omega^2\mathbf{r}, \tag{3.49}$$

where

$$\mathbf{v}' = v'\cos\alpha\,\hat{\mathbf{n}}_\theta + v'\sin\alpha\,\hat{\mathbf{n}}_\varphi \quad \text{and} \quad \vec{\omega} = \omega\hat{\mathbf{k}} = \omega\cos\theta\,\hat{\mathbf{n}}_r - \omega\sin\theta\,\hat{\mathbf{n}}_\theta.$$

By direct calculation, we obtain

$$\mathbf{N}_t = (2v'\sin\alpha + \omega r\sin\theta)m\omega\cos\theta\,\hat{\mathbf{n}}_\theta - 2mv'\omega\cos\alpha\cos\theta\,\hat{\mathbf{n}}_\varphi.$$

To determine $\mathbf{N}_\perp$, just note that

$$\mathbf{N}_\perp = \mathbf{N}_t - \frac{(\mathbf{N}_t\cdot\mathbf{v}')\mathbf{v}'}{v'^2}. \tag{3.50}$$

Thus, we arrive at the final result

$$\mathbf{N}_\perp = (2v' + \omega r\sin\alpha\sin\theta)(m\omega\sin\alpha\cos\theta\,\hat{\mathbf{n}}_\theta - m\omega\cos\alpha\cos\theta\,\hat{\mathbf{n}}_\varphi),$$

which, in terms of the latitude ξ, takes the form

$$\mathbf{N}_\perp = (2v' + \omega r\sin\alpha\cos\xi)(m\omega\sin\alpha\sin\xi\,\hat{\mathbf{n}}_\theta - m\omega\cos\alpha\sin\xi\,\hat{\mathbf{n}}_\varphi).$$

Note that the force acting on the rails vanishes at the equator and for $\alpha = 0$ (movement on a meridian), we obtain

$$\mathbf{N}_\perp = -2mv'\omega\sin\xi\,\hat{\mathbf{n}}_\varphi. \tag{3.51}$$

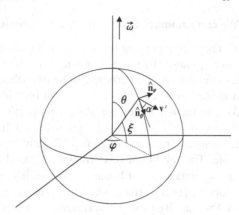

Fig. 3.20 The train in the frame of the Earth, which rotates with angular speed ω

Example 11. Show that the surface of water in a cylindrical bucket, which rotates about its axis of symmetry with constant angular speed ω, tends to a paraboloid of revolution. A paraboloid of revolution is obtained by rotating a parabola about the axis of symmetry.

Solution. On the plane xy, the equation of the parabola is given by $y = Bx^2 + Cx + D$. The problem can be solved in the rotating reference frame tied to a small portion of water (water element). We know that the surface of stagnant water, an inertial reference frame, is formed due to the gravitational force, so it is always perpendicular to the gravitational acceleration.

We already know that a non-inertial reference frame is locally equivalent to an inertial frame in a gravitational field (Principle of Equivalence). We will apply this principle to an element of water in the rotating reference frame. We consider this small portion of water in an inertial frame under the influence of the effective gravitational field given by $\mathbf{g}_{ef} = \mathbf{g} + \mathbf{a}_{cf}$, where $\mathbf{a}_{cf}$ is the centrifugal acceleration. Thus, the tangent line to the parabola must be perpendicular to the vector $\mathbf{g}_{ef} = \mathbf{g} + \mathbf{a}_{cf}$.

Indeed, $a_{cf} = v^2/x = \omega^2 x$, where x is the distance to the axis of rotation, so we can write $\mathbf{g}_{ef} = \omega^2 x \hat{\mathbf{i}} - g\hat{\mathbf{j}}$. The slope of $y(x)$ is defined by dy/dx, and it is easy to see that

$$\frac{dy}{dx} = \frac{\omega^2 x}{g}$$

is the requirement for $\mathbf{g}_{ef}$ to be perpendicular to the tangent line to the parabola. In this way we obtain

$$y = \frac{\omega^2}{2g}x^2 + \text{const},$$

that completes the proof. We will come back to this problem again in Chap. 10.

We can summarize the main characteristics of inertial forces as follows:

1) They appear during the accelerated motion of non-inertial reference and are not directly caused by interaction with some material objects.
2) The motion of a particle can be described in both types of frames, inertial and non-inertial. The inertial forces occur only in the latter case.
3) The inertial forces are always proportional to the particle masses. Therefore, they can be confused with gravitational forces. This "ambiguity" comprises a fundamental principle of General Relativity, called the Einstein's Equivalence Principle. The validity of this principle is based on equality between inertial mass and gravitational mass of bodies. This equality is a very important experimental result which has been verified with great precision, but is also a subject of further tests.
4) The validity of such characterization of inertial forces by means of the Equivalence Principle is restricted to a sufficiently small volume, for which the non-homogeneity of the gravitational and inertial forces can not be detected. For a better understanding, consider an extended body in free fall toward a planet. It is convenient to think about this experiment being done outside the atmosphere, so that only the gravitational force is present. For simplicity, assume that the center of the body (which may be, e.g., a spacecraft) approaches the center of the planet along a radial trajectory. If the ship is small, all bodies would fall within it the same way and the practical effect is that gravity could not be detected by the ship's passengers. In other words, without looking at the windows, the passengers of the sufficiently small space ship would never know whether they are very far from all gravitating bodies or are just in a free fall under the action of gravitational field.

Now let us consider that the size of the ship is not negligible. Each particle within it will go directly to the center of the planet, following a radial trajectory. In this case the astronauts have a chance to detect the difference in the motions of two test particles. The effect of gravity in this case can not be eliminated by choosing a frame in a free fall.

Finally, inertial and gravitational forces may compensate each other only in a small volume, which can be approximated to be point-like. Therefore, as discussed, the equivalence between inertia and gravitation is local and not global.

Additional Exercises

1. A block of mass M is shaped like a prism with the inclined plane forming an angle θ with the horizontal and can move on the horizontal surface without friction. Another small block of mass m can move on the inclined surface of the prism, also without friction (see Fig. 3.21). Find the acceleration of the prism using Newton's Laws.

Solution. Let us denote accelerations of the blocks of masses M and m, as $\mathbf{a}$ and $\mathbf{a}_1$, correspondingly. We also denote the acceleration of the block m with respect to the prism as $\mathbf{a}_1'$. Considering the axis OX to be horizontal and the axis OY vertical, we arrive at the relations

$$\mathbf{a}_1 = \mathbf{a} + \mathbf{a}_1', \qquad \mathbf{a}_1' = a_1'(\hat{\mathbf{i}}\cos\theta + \hat{\mathbf{j}}\sin\theta).$$

If N_1 and N_2 are normal forces acting respectively on the block of mass m and on the prism from the side of the horizontal surface, one can write

$$N_1 + m\mathbf{g} = m(\mathbf{a} + \mathbf{a}_1'), \qquad N_2 - N_1 + M\mathbf{g} = M\mathbf{a}.$$

Making a projection of these equations to the axes OX and OY, we come to the system of equations

$$N_1 \sin\theta = ma_1' \cos\theta - ma, \qquad -mg + N_1 \cos\theta = -ma_1' \sin\theta,$$
$$N_1 \sin\theta = Ma, \qquad N_2 - Mg - N_1 \cos\theta = 0.$$

Solving this system, one can easily obtain the answer

$$a = \frac{mg\sin\theta\cos\theta}{M + m\sin^2\theta}. \tag{3.52}$$

Observation. The same result can be obtained by solving the problem in the prism's reference frame (a non-inertial reference frame with acceleration **a**). In this frame, besides the forces N_1 and $m\mathbf{g}$, the inertial force $-m\mathbf{a}$ also acts on the block of mass m, which moves parallel to the inclined plane. In this approach, the solution is much simpler, and this enables us, without much effort, to introduce a coefficient of kinetic friction μ between the surface of the inclined plane and the block.

Solution. Considering the diagram of forces acting on the block of mass m, Newton's Second Law applied in the perpendicular direction to the surface of the inclined plane provides

$$N_1 = m(g\cos\theta - a\sin\theta).$$

At the same time, the horizontal component of Newton's Second Law applied to the prism provides

$$N_1 \sin\theta - N_1 \mu\cos\theta = Ma.$$

The elimination of N_1 from the system above enables one to write an equation for a and the solution has the form

$$a = \frac{mg\cos\theta(\sin\theta - \mu\cos\theta)}{M + m\sin\theta(\sin\theta - \mu\cos\theta)}.$$

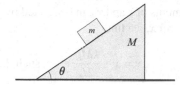

Fig. 3.21 Exercise 1

This equation reduces to the result (3.52) in the limit $\mu = 0$.

2. The system illustrated in Fig. 3.22 can move without friction and the pulleys have negligible mass and radius. Calculate the acceleration of the block $2m$ and tensions of the threads at the instant when the angle is α.

Solution. When the block in the middle descends the amount dx, the length of the thread in the middle part increases from l to $l+dl$. One can note that

$$(l+dl)^2 = (l \sin \alpha)^2 + (l \cos \alpha + dx)^2,$$

then $dl = dx \cos \alpha$. Therefore, if the acceleration of the block in the middle is given by a (downwards), the accelerations of the two other blocks are given by $a_1 = a \cos \alpha$ (upwards). According to Newton's Second Law, we have

$$mg - T = -ma_1 = -ma \cos \alpha, \qquad mg - T \cos \alpha = ma.$$

The solutions may be easily found in the form

$$a = \frac{1 - \cos \alpha}{1 + \cos^2 \alpha} g, \qquad T = \frac{1 + \cos \alpha}{1 + \cos^2 \alpha} mg.$$

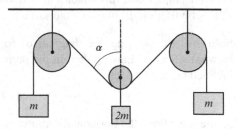

Fig. 3.22 Exercise 2

3. A sample of a thin chain has a length of a and stretches out by kFa when it is a subject to the force F. Considering a piece of the same wire of a non-stretched length L and mass M, what would be its length when it is hung by one of its ends?

Solution. Let x be the position of a point of the non-stretched wire, and y the position of the same point with the wire hanging. It is useful to define that $x = y = 0$ at the lowest point of the wire. Our purpose is to find the function $y(x)$. The force acting on the element dy is Mgx/L, hence one can note that the infinitesimal elements of x and y can be related by $dy = (1 + fx)dx$, where $f = kMg/L$. Integrating over x, we find

$$y = x + \frac{kMg}{2L} x^2, \quad \text{such that} \quad L_{hanged} = \int_0^L \frac{dy}{dx} dx = L\left(1 + \frac{kMg}{2}\right).$$

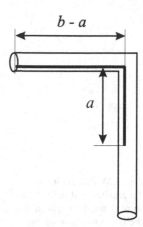

Fig. 3.23 Exercise 4

4. An inextensible homogeneous wire of mass M and length b is initially at rest in a tube in the manner illustrated in Fig. 3.23, such that the part of length a, where $a \lessdot b$, is loose in the vertical position and the rest of the wire has a horizontal position. There is no friction between the wire and the tube. Assume that the wire is free to descend and calculate the time required for the wire to descend to the completely vertical position. Explore the limit $a \to 0$ in the answer and explain the result.

Solution. At the moment when the length of the vertical part of the wire is x, the force which accelerates the wire has an absolute value Mgx/b. Therefore, x satisfies the differential equation

$$\ddot{x} = \lambda^2 x, \quad x(0) = a, \quad \text{where} \quad \lambda^2 = \frac{g}{b}.$$

Using the solution, $x = a \sinh \lambda t$, we find the requested time,

$$t_n = \sqrt{\frac{b}{g}} \operatorname{arcsinh}\left(\frac{b}{a}\right) = \sqrt{\frac{b}{g}} \ln\left(\frac{b}{a} + \sqrt{1 + \frac{b^2}{a^2}}\right). \tag{3.53}$$

In the limit $a \to 0$, the t_n becomes infinite, this reflects the fact that when $a = 0$, there is no motion.

Appendix

In this appendix we calculate the gravitational field generated by a spherical shell of negligible thickness and mass M and radius R, for the points both inside and outside the shell.

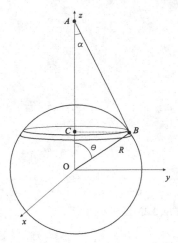

Fig. 3.24 Calculation of
the gravitational field at the
point A due to a spherical
shell. Contribution of an
infinitesimal element of mass

The areal mass density, ρ, can be obtained dividing the mass of the shell by its surface area,

$$\rho = \frac{M}{4\pi R^2}.$$

In order to find the gravitational force due to the shell that acts on a test particle of mass m, we use spherical coordinates, (r, φ, θ) and remember that the radius of the shell is fixed and equals R. Because of the symmetry of the spherical surface, there is no preferred direction to the position of the test particle. Then, without loss of generality, we assume the test particle to be placed at the OZ axis, such that its position is fully characterized by the coordinate z.

The mass in the infinitesimal area $d\theta d\varphi$ is given by

$$dM = \rho \, dS = \rho \cdot R^2 \sin\theta \, d\theta \, d\varphi = \frac{M}{4\pi} \sin\theta \, d\theta \, d\varphi.$$

The distance between the coordinate point (φ, θ) and the point with coordinates $(z, 0, 0)$ of the test particle is, according to Fig. 3.24,

$$AB = \sqrt{(z - R\cos\theta)^2 + R^2 \sin^2\theta} = \sqrt{z^2 - 2Rz\cos\theta + R^2}.$$

Therefore, according to Newton's law of gravitation, the differential element of the force has a magnitude given by

$$dF(\theta\,\varphi) = \frac{Gm \, dM}{AB^2} = \frac{GmM \sin\theta \, d\theta \, d\varphi}{4\pi \left[(z - R\cos\theta)^2 + R^2 \sin^2\theta\right]}. \tag{3.54}$$

The direction of this force makes an angle α with the axis OZ, as shown in Fig. 3.24. It can be noted immediately that

$$\sin \alpha = \frac{R\sin\theta}{AB} \quad \text{and} \quad \cos\alpha = \frac{z - R\cos\theta}{AB}. \tag{3.55}$$

The total force acting on a test particle can be obtained by integration on θ and φ. It is more useful to integrate first over φ. In this case, the resultant force (still infinitesimal) will be

$$d\mathbf{F}(\theta) = \int_0^{2\pi} d\mathbf{F}(d\theta\,d\varphi) = \int_0^{2\pi} \frac{d\mathbf{F}}{d\varphi}\,d\varphi,$$

where the integration is over φ. By reasons of symmetry, it is clear that the last differential element of force is directed parallel to the axis OZ. So when we perform the last integration, we can multiply (3.54) by $\cos\alpha$ from (3.55).

Thus, we obtain

$$dF(\theta) = d\mathbf{F}(\theta) \cdot \hat{\mathbf{k}},$$

where the magnitude of the force corresponds to the contribution of the area between θ and $\theta + d\theta$,

$$dF(\theta) = -\frac{Gm\,dM}{AB^2}\cdot\cos\alpha = -\frac{2\pi R^2 \rho mG\,(z - R\cos\theta)\sin\theta\,d\theta}{\left[z^2 + R^2 - 2Rz\cos\theta\right]^{3/2}}. \tag{3.56}$$

For the total force, one has to integrate

$$\int_0^{\pi} \frac{dF(\theta)}{d\theta}\,d\theta = -2\pi R^2 \rho\,mG\int_0^{\pi} \frac{(z - R\cos\theta)\sin\theta\,d\theta}{\left[z^2 + R^2 - 2Rz\cos\theta\right]^{3/2}}.$$

This integral is relatively difficult, but we can simplify it if taking into account that

$$\frac{z - R\cos\theta}{\left(z^2 + R^2 - 2Rz\cos\theta\right)^{3/2}} = \frac{d}{dz}\left(\frac{1}{\sqrt{z^2 + R^2 - 2Rz\cos\theta}}\right).$$

The integration over the angle θ and the derivative with respect to z are commutative operations, i.e., the result does not depend on the order of their execution. Now we just calculate the integral

$$-\int_0^{\pi} \frac{dU(z,\theta)}{d\theta}\,d\theta,$$

where[6]

$$dU(z,\theta) = -2\pi R^2 \rho mG\frac{\sin\theta\,d\theta}{\sqrt{z^2 + R^2 - 2zR\cos\theta}}, \tag{3.57}$$

---

[6] The intermediate function $U(z)$ is called *potential energy* of the test mass in the gravitational field generated by the spherical shell. Later on, we discuss the properties of potential energy with more details.

and by taking the derivative d/dz of the result of this integration, $U(z) = \int_0^\pi dU(z, \theta)$,

$$F(z) = -\frac{dU}{dz}.$$

The integration of the Eq. (3.57) can be performed in the following way. We have

$$-U(z) = \int_0^\pi 2\pi R^2 \rho Gm \frac{\sin\theta\, d\theta}{\sqrt{z^2 + R^2 - 2zR\cos\theta}}. \tag{3.58}$$

Consider the change of variables

$$x = z^2 + R^2 - 2zR\cos\theta, \qquad dx = 2zR\sin\theta\, d\theta,$$
$$x(0) = (R - z)^2, \qquad x(\pi) = (R + z)^2.$$

Then we have

$$-U(z) = \frac{2\pi R^2 \rho Gm}{2zR} \int_{(R-z)^2}^{(R+z)^2} \frac{dx}{x^{1/2}} = \frac{\pi R\rho Gm}{z} \cdot 2x^{1/2} \Big|_{(R-z)^2}^{(R+z)^2}$$

$$= \frac{2\pi R\rho Gm}{z} \left\{ |R + z| - |R - z| \right\}. \tag{3.59}$$

The final formula has a content which is quite different for the points inside and outside the spherical shell, that is for the cases $z < R$ and $z > R$.

For $z < R$ we find $|R + z| - |R - z| = R + z - R + z = 2z$, so that

$$U(z) = -4\pi R\rho Gm = \text{const}.$$

Obviously, after taking the derivative with respect to z, we obtain zero, $F(z) = 0$. Thus, inside the spherical shell, the gravitational field is null.

For $z > R$ we have $|R + z| - |R - z| = R + z + R - z = 2R$. Thus,

$$U(z) = -\frac{4\pi R^2 \rho\, Gm}{z} = -\frac{GmM}{z}, \tag{3.60}$$

where we have used the equality $4\pi R^2 \rho = M$. This is the same result as in the case where the whole mass is concentrated in the center of the spherical shell. Thus, outside the shell the gravitational force is

$$F(z) = -\frac{GmM}{z^2}.$$

In the case of a solid body having a spherical symmetry in the mass distribution, we can divide it into infinitely thin spherical shells (see Fig. 3.25). To calculate the gravitational field in the point A inside the body, one has to take into consideration only the mass within a sphere of the radius r, where $r = |\overrightarrow{AO}|$.

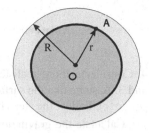

Fig. 3.25 A sphere obtained as an integral of elementary spherical shells

Exercises

1. The people from the world X have learned to dig holes in their land. They made a straight and slim tunnel through the center of their (spherically symmetric) planet and found that the density of the planet is always the same, regardless of the distance to the center r. How does the acceleration of gravity depend on the depth of the tunnel h?

Answer: $g(r) = gr/R$, where g is the gravitational acceleration on the surface and R is the radius of the planet X (so that $r = R - h$).

2. A galaxy can be approximated by a circular disk of radius R and mass M. Assuming that the mass of this galaxy is distributed in a nearly homogeneous way in the disc plane, calculate the gravitational force that the galaxy exerts on a test mass m, located on the axis of symmetry of the disk at the distance z from its center. Also calculate the gravitational field, $g(z) = F/m$. Note that the gravitational field here is the gravitational acceleration $g(z)$ (Fig. 3.26).

$$\textbf{Answer:} \qquad g = \frac{2GM}{R^2}\left[1 - \frac{z}{\sqrt{z^2 + R^2}}\right].$$

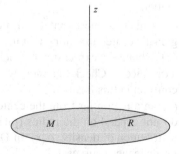

Fig. 3.26 A circular disk representing a galaxy

3. Consider the same model of the galaxy as in the previous problem, but this time calculate the gravitational field $g(z)$, in the plane of the galaxy, at a point at a distance z from its center. The answer should have the form of an integral.

Answer: $g = \dfrac{GM}{\pi R^2} \displaystyle\int_0^{2\pi}\!\!\int_0^R \dfrac{(z - r\cos\varphi)\, r\, dr\, d\varphi}{\left(z^2 + r^2 - 2zr\cos\varphi\right)^{3/2}}.$

4. Consider a more realistic model of a galaxy in comparison with Exercises 2 and 3. Assume that the surface mass density varies with the distance r to the center: $\rho = \rho(r)$. Follow the steps below:

(a) Calculate the gravitational field, $g(x)$, in a point in the plane of the galaxy, at the distance x from the center. The answer should have the form of an integral representation.

(b) A star follows a circular orbit around the center of the galaxy with velocity $v(x)$. Determine $v(x)$. **Answers:**

$$\text{(a)}\quad g(x) = G \int_0^{2\pi}\!\!\int_0^R \frac{(x - r\cos\varphi)\,\rho(r)\, r\, dr\, d\varphi}{\left(x^2 + r^2 - 2xr\cos\varphi\right)^{3/2}};$$

$$\text{(b)}\quad v(x) = \sqrt{g(x)\,x}.$$

5. Problem of Dark Matter. According to astronomical data, the dependence of the rotational speeds of stars in spiral galaxies and the bending radius of their circular paths is given by a complicated formula, but it can be approximated, in a region near the center of the spiral galaxy, by a linear law, $v = \text{const} \cdot r$, where r is the distance to the center of the galaxy. One can show that the mass of visible matter can not provide this law. A most convincing model for this phenomenon implies that, in addition to the luminous (visible) matter distributed approximately in the plane of the galaxy, there is a so-called halo of dark matter, which represents an almost spherically symmetric and homogeneously distributed mass. The total mass of the dark matter is dominant in defining the law of rotation of matter, including stars.

Consider a simplified model of the galaxy. The task is to show that the law $v = \text{const} \cdot r$ corresponds to the spherically symmetric and homogeneous distribution of matter.

(a) Calculate the gravitational field, $g(x)$, in a point at the distance r from the galactic center, assuming that the mass of the halo is M and that its radius is R.

(b) Obtain the expression for the speed $v(x)$ of a star orbiting around the center, at a distance r. Check that the law $v = \text{const} \cdot r$ is valid and calculate the value of the constant in this formula.

(c) In a region far from the center of the galaxy, the velocities of the stars almost do not depend on the radius (plane curves of rotation). Calculate the symmetrical distribution of density $\rho(r)$ of Dark Matter in the halo of the galaxy that is able to produce these curves.

Answers:

$$(a) \quad g(r) = \frac{GM}{R^3} r; \qquad (b) \quad v(r) = \sqrt{\frac{GM}{R^3}} r; \qquad (c) \quad \rho(r) = \frac{v_0^2}{4\pi G r^2}.$$

6. A thin rope is attached to a boat and wrapped around the column of the port n times, according to Fig. 3.27. Assume that the friction coefficient between the rope and the column is given by $\mu = 0.25$. Find the value of n for which a 3 year old boy can handle the boat. Consider that the child can develop a force 20 N and that 20,000 N are needed to hold the boat.

Solution. We can easily build the differential equation for the tension force, F, as a function of the angle φ, defined as the angle in the cross section of the column between the point where 20,000 N is applied and a generic point in the rope. Taking into account also the friction force in an infinitesimal part of the rope, we get

$$dF = -2\mu F \sin\frac{d\varphi}{2} \approx -\mu F d\varphi, \qquad F(\varphi = 0) = F_0 = 20,000\,\text{N}.$$

The solution of this equation has the form $F = F_0 \exp(-\mu\,\varphi)$. For the number of necessary turns around the column, we find $n = \frac{3\ln 10}{2\pi\mu} \approx 4.4$.

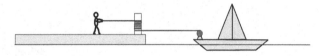

Fig. 3.27 Illustration of Exercise 6

Chapter 4
Conservation of Momentum

Abstract We shall treat momentum conservation as a consequence of Newtonian dynamics, despite the fact that it can be deduced from the symmetry considerations (space homogeneity property) in a more sophisticated formalism. The center of mass of a system of particles is a related notion which will be introduced here. We are going to show how the problem of the system of two interacting particles can be reduced to the dynamics of a single particle with reduced mass in a central field.

4.1 Linear Momentum

Consider a system of particles which are subject to Newton's laws, in an inertial reference frame. We can use the second Newton law to write a system of differential equations for the radius-vectors of the particles in the form

$$m_i \mathbf{a}_i = m_i \ddot{\mathbf{r}}_i = \mathbf{F}_i(t; \mathbf{r}_1, \mathbf{r}_2, \ldots, \mathbf{r}_N; \mathbf{v}_1, \ldots \mathbf{v}_N), \qquad (4.1)$$

where N is the number of particles, m_i, $\mathbf{r}_i(t)$ and $\mathbf{v}_i = \dot{\mathbf{r}}_i$ are the mass, the position vector and the velocity of i-th particle ($i = 1, 2, \ldots, N$). The vector $\mathbf{F}_i$ is the force exerted on the i-th particle. According to the principle of superposition,

$$\mathbf{F}_i = \sum_{j \neq i} \mathbf{F}_{ji} + \mathbf{F}_i^{ext}, \qquad (4.2)$$

where $\mathbf{F}_i^{ext}$ is the external force and $\mathbf{F}_{ji}$ is the force that the j-th particle exerts on the i-th particle. Here we assume that $j = 1, 2, \ldots, i-1, i+1, \ldots, N$. As usual, we assume that the forces depend on the position vectors and velocities, but not on the accelerations and their time derivatives. This property is valid for all the forces discussed previously.[1]

---

[1] A remarkable exception is the electromagnetic force between two moving charges [23]. The completely consistent consideration of this type of interaction can be done only in the scope of the

I.L. Shapiro and G. de Berredo-Peixoto, *Lecture Notes on Newtonian Mechanics*, Undergraduate Lecture Notes in Physics, DOI 10.1007/978-1-4614-7825-6_4, © Springer Science+Business Media, LLC 2013

Let us define a new concept, the linear momentum, or simply momentum. For a particle of mass m and velocity $\mathbf{v}$, the momentum is defined by

$$\mathbf{p} = m\mathbf{v}. \tag{4.3}$$

According to the last relation, the momentum of the particle is a vector that has the same direction as its velocity. The momentum of a system of particles is defined as the sum of the momenta of each constituent particle, $m_1, m_2, \ldots, m_N$, i.e.,

$$\mathbf{P} = \mathbf{p}_1 + \mathbf{p}_2 + \ldots + \mathbf{p}_N.$$

Considering that the mass of each particle is constant, one can rewrite the Eq. (4.1) as

$$\dot{\mathbf{p}}_i = \mathbf{F}_i = \sum_{j \neq i} \mathbf{F}_{ji} + \mathbf{F}_i^{ext}, \tag{4.4}$$

where the total force acting on the particle is written in the form (4.2). The last equation represents Newton's Second Law in terms of momenta.

The law of conservation of linear momentum is valid for *closed systems*, i.e., systems free of external forces, with $\mathbf{F}_i^{ext} = 0$. The formulation of the law of conservation of momentum is as follows:

For a closed system constituted of N particles of masses $m_1, m_2, \ldots, m_N$, its total momentum, $\mathbf{P}$, does not depend on time, i.e., $\dot{\mathbf{P}} = 0$.

For a better understanding of this law, before a formal proof, let us consider the two examples, namely one of a free particle and another one of a closed system of two particles. It is interesting that the first example was known already to Aristotle (382–322 BC). In his famous book *Physics* he presented arguments against the existence of vacuum as an absolutely empty place.[2] The consideration starts from the assumption of homogeneity of space and proceeded with the following logic (which we present in modern terms):

For a free particle there are no external forces and the unique feature of the movement is the speed. As time goes on, only the position of the particle changes, but as far as all points in space are equal, the speed can not change and therefore has to remain constant.

We can achieve the same result using equations. The momentum of a free particle is given by $\mathbf{p} = m\mathbf{v}$. Using Newton's Second Law, we get

$$\frac{d\mathbf{p}}{dt} = m\frac{d\mathbf{v}}{dt} = m\mathbf{a} = \mathbf{F} = 0,$$

because the particle is free.

Theory of Relativity (Relativistic Electrodynamics). Therefore, the complete discussion of electromagnetic interactions is not possible in the present book.

[2] This is, actually, not infinitely far from the modern quantum understanding of vacuum.

As a next step, let us consider two particles of masses m_1 and m_2, with velocities v_1 and v_2. The momentum of the system is given by

$$\mathbf{P} = m_1 \mathbf{v}_1 + m_2 \mathbf{v}_2 .$$

According to Newton's third law, the forces exerted by the particles satisfy the relation $\mathbf{F}_{12} = -\mathbf{F}_{21}$ (see Fig. 3.3). Furthermore, since the system is closed, there are no external forces, i.e., $\mathbf{F}_1^{ext} = \mathbf{F}_2^{ext} = 0$. Using the Eq. (4.4), we found the following equation for the linear momentum of the system:

$$\frac{d\mathbf{P}}{dt} = m_1 \frac{d\mathbf{v}_1}{dt} + m_2 \frac{d\mathbf{v}_2}{dt} = m_1 \mathbf{a}_1 + m_2 \mathbf{a}_2 = \mathbf{F}_{21} + \mathbf{F}_{12} = 0, \qquad (4.5)$$

from what we conclude that $\mathbf{P} = \text{const.}$

The most important aspect in the above consideration is the absence of external forces. The next step is a formal statement about the law of conservation of linear momentum for a closed system of N particles. The total momentum has the form

$$\mathbf{P} = m_1 \mathbf{v}_1 + m_2 \mathbf{v}_2 + \ldots + m_N \mathbf{v}_N = \sum_{i=1}^{N} m_i \mathbf{v}_i .$$

The resultant force exerted on the i-th particle is given by

$$\mathbf{F}_i = \sum_{j=1, j \neq i}^{N} \mathbf{F}_{ji} .$$

Newton's Second Law for the system can be written as

$$\frac{d\mathbf{P}}{dt} = \sum_{i=1}^{N} m_i \mathbf{a}_i = \sum_{i=1}^{N} \mathbf{F}_i = \sum_{i=1}^{N} \sum_{j \neq i, j=1}^{N} \mathbf{F}_{ji} .$$

The internal sum covers all particles with $j \neq i$, because the i-th particle does not exert force on itself. Considering Newton's Third Law, $\mathbf{F}_{ij} = -\mathbf{F}_{ji}$, and regrouping the total sum of forces in a sum of pairs action-and-reaction, we observe that the total sum will be zero, since the sum of each such pair vanishes independently. Thus, we obtain $\mathbf{P} = \text{constant.}$

It is worthwhile to make an observation that may be important for the practical application of the law of conservation of momentum. It is easy to see that the condition $\mathbf{F}_i^{ext} = 0$ can be reduced to a much weaker requirement, namely

$$\sum_{i=1}^{N} \mathbf{F}_i^{ext} = 0. \qquad (4.6)$$

If external forces are present, but the resulting force (4.6) is zero, the momentum of the system is conserved in the same manner as in the case of a closed system. Furthermore, vector equality (4.6) can be decomposed into three equalities for the components F_x, F_y, F_z for the sum of external forces. If, for example, the component F_x is null, the linear momentum component P_x of the system will be conserved, independent of whether the same is true or not for other two components.

We have demonstrated the law of conservation of linear momentum for a closed system of N particles. In general, a quantity that does not change during the motion is called *integral of motion* or *constant of motion*. Using this terms, we have shown that the total momentum of a closed system is an integral of motion.

This result can be immediately generalized. Any body can be seen as a system of particles (e.g., atoms or molecules) interacting with each other. So the statement also applies to any body, since the absence or the cancelation of the external forces is assured.

The next important observation concerns a mass exchange which may occur between the particles, e.g., in a process

$$(m_1 + \Delta m) + m_2 \; \to \; m_1 + (m_2 + \Delta m) \,.$$

This kind of mass exchange, however, can be always regarded as a process in a system of three masses, m_1, m_2 and Δm, and therefore it does not deserve a separate consideration. The momentum is preserved because there are no external forces in this system.

To conclude our discussion about conservation of momentum, this conservation law is one of the most general ones in Physics. It holds even in the much more complicated cases, for example in relativistic and quantum theories, while the mass is not necessarily a constant of motion and the definition of momentum of a particle is modified.

One can understand the existence of the integrals of motion in a more general framework. It is natural to ask the following question: what are the other possible constants of motion, besides the linear momentum? To answer this question we need to go back to the equations of motion, Eq. (4.1). To complete the definition of the dynamical problem, we need to define a set of initial data, e.g., positions and velocities of all the particles at the initial instant of time t_0, $\mathbf{r}_i(t_0)$ and $\mathbf{v}_i(t_0)$.[3]

As an example, consider the simple case of a free particle, when there are six constants of integration in the equations of Newton. When the particle is free, three of these constants represent the components of momentum, which are constants of motion. The other three constants can be, for example, the coordinates of the particle at the instant $t = 0$.

For the system of N particles, without constraints, each of them has three components of radius vector and three components of velocity. Then, the total number of constants of integration is $6N$. So, the integration of the system (4.1) generates a set of $6N$ constants of integration, C_k $(k = 1, 2, \ldots, 6N)$. Then, the general solution of (4.1) has the form

$$\mathbf{r}_i = \mathbf{r}_i(t, C_k) \,, \quad \text{where} \quad k = 1, 2, \ldots, 6N \,. \tag{4.7}$$

---

[3] The equations set and initial data is called Cauchy's problem, in honor of nineteenth-century French mathematician, A.-L. Cauchy, who made great contributions in Mathematics and Mechanics.

The values of C_k can be defined using the initial data $\mathbf{r}_i(t_0)$ and $\mathbf{v}_i(t_0)$. We can solve (4.7) for the constants and obtain inverse relations

$$C_k = C_k[t, \mathbf{r}_i(t), \dot{\mathbf{r}}_i(t)] = \text{const.} \tag{4.8}$$

These expressions are nothing but the combinations of coordinates and velocities of the system which remain constant during the motion. In other words, those are constants of motion. One can see that the system of N particles always has $2N$ such constants, or $2N$ integrals of motion. For a system of many particles, it is a large number and, of course, most of these constants may not have physical interpretation. On the other hand, some constants of motion are related to the symmetries of space and time. These integrals of motion are very important, because they reflect fundamental properties of space and time. An example is the linear momentum which is related to the homogeneity of space. In the course of Analytical Mechanics, these issues can be discussed with somewhat greater clarity and generality. At the present level of introductory course we have to rely much more on our intuition, and in the case of conservation of momentum this approach has a long history, starting from Aristotle.

Let us consider some examples of application of the conservation of momentum in mechanics.

Example 1. Totally inelastic collision of two bodies of masses m and M.

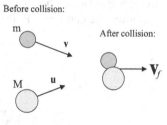

Fig. 4.1 Totally inelastic
collision

Suppose that before collision the velocities of the bodies were given by $\mathbf{v}$ and $\mathbf{u}$ (as shown in Fig. 4.1), and after they collide they move together with velocity equal to $\mathbf{V}_f$.

Our goal is to calculate $\mathbf{V}_f$. It is remarkable that this problem can be easily solved without knowledge of the details of the collision (which are usually quite complex). The system is closed because there are no external forces. Using the conservation of momentum, we find $\mathbf{P}_i = \mathbf{P}_f$, i.e.,

$$m\mathbf{v} + M\mathbf{u} = (m+M)\mathbf{V}_f.$$

This means, of course,

$$\mathbf{V}_f = \frac{m\mathbf{v} + M\mathbf{u}}{m+M}.$$

Notice that when we take $\mathbf{u} = 0$,

$$\mathbf{V}_f = \frac{m\mathbf{v}}{m+M}.$$

Example 2. The same problem can be solved for a completely inelastic collision of many bodies. If the masses involved are given by $m_1, m_2, \ldots m_N$ and the velocities are $\mathbf{v}_1, \mathbf{v}_2, \ldots \mathbf{v}_N$, the momentum of the system before the collision is

$$\mathbf{P} = m_1\mathbf{v}_1 + m_2\mathbf{v}_2 + \ldots + m_n\mathbf{v}_N. \tag{4.9}$$

As far as the system is closed, the momentum will remain the same after the collision, when the particles join together and form a single body of mass

$$M = m_1 + m_2 + \ldots + m_N.$$

For the speed of this body, $\mathbf{V}$, we have the relation

$$M\mathbf{V} = \mathbf{P} = m_1\mathbf{v}_1 + m_2\mathbf{v}_2 + \ldots + m_n\mathbf{v}_N,$$

i.e.,

$$\mathbf{V} = \frac{m_1\mathbf{v}_1 + m_2\mathbf{v}_2 + \ldots + m_N\mathbf{v}_n}{m_1 + m_2 + \ldots + m_N}. \tag{4.10}$$

The above result, Eq. (4.10), remains valid even in collisions or other forms of interaction between the particles. The total momentum of the system is a constant of motion. We can write $\mathbf{P}$ in another form, noting that

$$\mathbf{v}_i = \frac{d\mathbf{r}_i}{dt}$$

for each particle labeled by i. In this case,

$$\mathbf{P} = \sum_{i=1}^{n} m_i\mathbf{v}_i = \sum_{i=1}^{N} m_i\frac{d\mathbf{r}_i}{dt} = \sum_{i=1}^{N} \frac{d}{dt}(m_i\mathbf{r}_i) = \frac{d}{dt}\sum_{i=1}^{N} m_i\mathbf{r}_i.$$

If we divide the last expression for the total mass M, we obtain a new vector quantity, $\mathbf{V}$, called velocity of the center of mass of the system,

$$\mathbf{V} = \frac{\mathbf{P}}{M} = \frac{1}{M}\sum_{i=1}^{N} m_i\mathbf{v}_i = \frac{d\mathbf{R}}{dt}. \tag{4.11}$$

In the last equation, we also define the position vector for the center of mass of the particle system,

$$\mathbf{R} = \frac{1}{M}\sum_{i=1}^{N} m_i\mathbf{r}_i. \tag{4.12}$$

For any closed system, as a consequence of conservation of linear momentum, we have

$$\dot{\mathbf{R}} = \mathbf{V} = \text{const.}, \tag{4.13}$$

i.e., the movement of the center of mass of a closed system has exactly the same motion, as the one of a free particle. The main difference between a free particle and a closed system of many particles is that for these particles there may exist an inner dynamics, in particular caused by forces between the particles. The existence of the center of mass with the property (4.13) is a manifestation of the law of conservation of linear momentum. The internal dynamics cannot influence the motion of the center of mass.

Finally, we mention that the notion of center of mass (4.12) can be defined and becomes fruitful also in the general case, when the system is not closed. In this case, one can consider the motion of the entire system in terms of the resultant force and mass. The condition (4.13) indicates the particular case, in which the resultant force acting on the system is null.

To describe the inner dynamics of a system of particles, it is desirable to introduce a new reference frame (C), whose origin, O_C, coincides with the center of mass of a particle system, i.e., $OO_C = \mathbf{R}$, where O is the origin of an external inertial frame reference or "laboratory reference frame" (L). Consider a closed system. In this case, taking into account that $\mathbf{V} = \dot{\mathbf{R}} = const$, we note that the reference frame (C) is also inertial.

The transformation between the reference frames (L) and (C) is given by:

$$\mathbf{r}_i = \mathbf{r}'_i + \mathbf{R}, \quad \text{and so} \quad \mathbf{v}_i = \mathbf{v}'_i + \mathbf{V}, \tag{4.14}$$

where $\mathbf{r}_i$ and $\mathbf{v}_i$ are the radius vector and the speed of the i-th particle in the reference (L), and $\mathbf{r}'_i$ and $\mathbf{v}'_i$ are similar quantities in the center of mass reference frame. The use of the reference system (C) may be very useful in different problems of Classical Mechanics and also in other parts of Physics, including Relativistic Quantum Mechanics and the theory of Elementary Particles (which is, simultaneously, both quantum and relativistic theory). In this book we will use the reference system (C) several times.

Example 3. We consider the motion of the jet propulsion of a rocket of mass M. The mass of fuel (along with the oxygen that participates in the combustion) is given by m and the output speed of the fuel rocket propellant has the value c. Calculate the velocity of the rocket as a function of the amount of remaining fuel and also its final velocity. Suppose that there is no air resistance force outside and that the movement of the rocket is rectilinear.

Solution. Considering the output of a small portion of the fuel, $-d\mu$, at the moment in which the amount of fuel in the tank of the rocket is μ and the total mass of the rocket is $\mu + M$, one can write the law of conservation of momentum in the form (note that the velocity of the propellant in the inertial reference frame is $v - c$)

$$(M + \mu)v = (M + \mu + d\mu)(v + dv) - (v - c)d\mu,$$

with the initial condition $\mu(v = 0) = m$. Cancelling identical terms on both sides of the equation, taking into account only the first order infinitesimal and separating the variables, we obtain

$$(M + \mu)dv = -cd\mu, \qquad \frac{d\mu}{\mu + M} = -\frac{dv}{c}, \qquad \ln\left|\frac{\mu + M}{C_1}\right| = -\frac{v}{c}.$$

Using the initial condition, we obtain the value of the constant of integration, $C_1 = m + M$. Finally,

$$v(\mu) = c \ln\left(\frac{m + M}{\mu + M}\right). \tag{4.15}$$

At the end of the acceleration process there is $\mu = 0$ and thus the maximum speed which is possible for that type of rocket is given by (Tsiolkovsky formula)

$$v_{max} = c \ln\left(1 + \frac{m}{M}\right). \tag{4.16}$$

Finally, let us consider non-closed systems (subject to some external influences) and discuss how the force exerted on a body changes its momentum. Consider the case when an external force $\mathbf{F}(t)$ is exerted on a body of mass m. Suppose that at an initial instant t_1 the body has the momentum $\mathbf{P}_1$. What will be the momentum at the instant $t_2 > t_1$?

To solve this problem, consider Newton's Second Law,

$$\mathbf{F} = m\mathbf{a} = m\frac{d\mathbf{V}}{dt} = \frac{d(m\mathbf{V})}{dt} = \frac{d\mathbf{P}}{dt},$$

where $\mathbf{V}$ is the velocity of the mass center of the body and $\mathbf{P}$ is its momentum at the instant of time t. Integrating the last relation between the time instants t_1 and t_2, we obtain

$$\int_{t_1}^{t_2} \mathbf{F}(t)dt = \int_{t_1}^{t_2} \frac{d\mathbf{P}}{dt}dt = \int_{t_1}^{t_2} d\mathbf{P}(t) = \mathbf{P}(t_2) - \mathbf{P}(t_1) = \Delta\mathbf{P}. \tag{4.17}$$

The vector quantity

$$\Delta\mathbf{P} = \int_{t_1}^{t_2} \mathbf{F}(t)dt$$

is called the impulse of the force $\mathbf{F}$. As we have seen, the change of momentum of the body in the range between t_2 and t_1 is equal to the impulse of force $\mathbf{F}(t)$ exerted on this body. If more than one force acts on a given body, it is necessary to integrate the resultant force. By the principle of superposition, the result is the same compared to the one that would be obtained by summing up the individual impulses of each force.

Exercises

1. What is the total momentum of a system of particles in the frame of its center of mass?

2. An elephant of mass m is initially at one end of a train flatcar of length L and mass M. The elephant moves (calmly) to the other end of the flatcar. Calculate the displacement of the flatcar, y, in relation to the rails. There is no friction between the wheels and rails.

Solution. There are no external forces along the rails, hence the position of the center of mass of the system remains constant. By using this condition, we obtain the equation

$$\frac{ML}{2} = M\left(\frac{L}{2} - y\right) + m(L - y),$$

from where one can easily obtain $y = mL/(m+M)$.

3. In the previous problem, at a given moment, the elephant reached a speed v with respect to the flatcar. Calculate its speed with respect to the rails at that instant. Taking into account that

$$\int_0^t v(t')dt' = L,$$

where t is the time for the elephant to walk along the flatcar, check the result of the previous problem.

4. An elephant, preparing to be a circus performer, jumped from a flatcar *(A)* of the train to another one, *(B)*. The flatcars are not linked. Initially, the elephant and both flatcars are at rest. At the instant just before the jump, the speed of the elephant with respect to the flatcar *(A)* was v_0, and after the (successful) jump, the elephant stopped to rest at the other flatcar. Calculate the final speeds of the two flatcars. The masses of the elephant and flatcars are given in Exercise 2.

Answer: $v_A = -\frac{m v_0}{m+M}$ and $v_B = \frac{mM v_0}{(m+M)^2}$.

5. Solve the problem of a completely inelastic collision by using the transformation to the reference system (C).

6. A particle of mass m and electric charge q enters a region subject to a magnetic field, uniform and constant, $\mathbf{B} = B\hat{\mathbf{k}}$. The initial velocity of the particle at the moment $t = 0$ is given by

$$\mathbf{v}_0 = \frac{v_0}{\sqrt{2}} \left(\hat{\mathbf{i}} + \hat{\mathbf{k}}\right). \tag{4.18}$$

All forces besides the magnetic one can be considered weak and effectively ignored. Perform the following tasks:

(a) Using the Eq. (3.9), write Newton's Second Law for this particle.

(b) Solve the equation and find the dynamic equation of motion and the trajectory.

(c) Write the vector momentum of the particle in terms of velocity components of the particle and determine its modulus. Check which one of the components of linear momentum in Cartesian coordinates is a constant of motion and explain this result qualitatively.

(d) Calculate the impulse of the magnetic force, $\Delta P(\Delta t)$ during a time interval Δt. Explain the existence of the period T, such that $\Delta P(T) = 0$. Discuss the dependence between T and the velocity v_0. Without doing calculations, obtain T for the case of the initial velocity given by

$$\mathbf{v_0} = v_0 \left(\hat{\mathbf{i}} \cos \alpha + \hat{\mathbf{j}} \cos \beta + \hat{\mathbf{k}} \cos \gamma \right), \qquad (4.19)$$

$$\text{where} \quad \cos^2 \alpha + \cos^2 \beta + \cos^2 \gamma = 1.$$

(e) Try to approach the problem of determining the period T by using dimensional arguments. Compare the output with the exact result.

(f) For the initial condition (4.18), calculate the radius of curvature of the trajectory, R, and find its relation to the period T. Without doing calculations, establish a relation analogous to the general case (4.19).

Answers and solutions:

$$(a) \quad \dot{v}_x = \omega v_y, \quad \dot{v}_y = -\omega v_x, \quad \omega = \frac{qB}{m}.$$

The last expression is called the cyclotron frequency. The remarkable feature of this frequency is its universality. This means it does not depend on the speed of the particle and on the angle between the initial velocity and direction of the magnetic field.

$$(b) \quad v_x = \frac{v_0}{\sqrt{2}} \cos \omega t, \quad v_y = -\frac{v_0}{\sqrt{2}} \sin \omega t, \quad v_z = \frac{v_0}{\sqrt{2}}.$$

(c) The trajectory is a spiral line. The motion is described by

$$x = x_0 + \frac{v_0 \sin \omega t}{\sqrt{2}\omega}, \quad y = y_0 + \frac{v_0 \cos \omega t}{\sqrt{2}\omega}, \quad z = z_0 + \frac{v_0 t}{\sqrt{2}}.$$

(d) The impulse of the magnetic force is

$$\Delta P = \int_t^{t+\Delta t} m\dot{v} \, dt',$$

it is periodic function with the period $T = 2\pi/\omega$. In the case (4.19), the frequency is given by $\omega = qB/m$.

(e) Let us assume that the frequency does not depend on the velocity v_0. Then we need to figure out how the frequency of motion depends on the three quantities q, B, m. Let us assume that the formula of our interest has the form

$$\omega = C \cdot B^{\alpha} q^{\beta} m^{\gamma},$$

where α, β, γ are the coefficients which we intend to learn and C is a dimensionless constant.

Using the dimensions of the quantities of our interest in the Gaussian system of units,

$$[\omega] = s^{-1}, \quad [q] = g^{1/2} cm^{3/2} s^{-1}, \quad [B] = g^{1/2} cm^{-3/2}, \quad [m] = g,$$

we arrive at the linear relations for α, β, γ,

$$
\begin{array}{ll}
\text{gram} & \dfrac{\alpha}{2} + \dfrac{\beta}{2} + \gamma = 0, \\[2mm]
\text{sec} & \beta = 1, \\[2mm]
\text{cm} & \dfrac{3\alpha}{2} + \dfrac{3\beta}{2} = 0
\end{array}
\tag{4.20}
$$

Solving this system we arrive at the unique solution, $\alpha = \beta = 1$, $\gamma = -1$, and therefore we arrive at the formula $\omega = CqB/m$, with $T = 2\pi/\omega$. Let us note that the dimensional method is not capable of providing the coefficient C and in this sense it is not precise. Moreover, the choice of physical assumptions here is especially important. For example, if we do not exclude the dependence on v_0 beforehand, there is no way to solve the dimensional system in a unique way, because the number of equations will be smaller than the number of variables $\alpha, \beta, \gamma, \ldots$.

$$(f) \qquad R = \frac{v_0}{\omega \sin \gamma} = \frac{m v_0}{q B \sin \gamma}.$$

4.2 The Problem of Two Bodies

The problem of two bodies has special importance in Classical Mechanics and all related areas, such as Quantum Mechanics, General Relativity and others. Our task is to consider the motion of a closed system of two point particles of masses m_1 and m_2. The radius-vectors of the particles are $\mathbf{r}_1$ and $\mathbf{r}_2$, respectively. As far as the system is closed, there is no external force, we also assume that the particles interact through a central force as

$$\mathbf{F}_{12} = -\mathbf{F}_{21} = \mathbf{F},$$

where

$$\mathbf{F} = F \cdot \hat{\mathbf{r}}, \qquad \hat{\mathbf{r}} = \frac{\mathbf{r}}{r}, \qquad \text{and} \qquad \mathbf{r} = \mathbf{r}_2 - \mathbf{r}_1. \tag{4.21}$$

The situation is illustrated in Fig. 4.2.

A relevant observation is that the central force is the only kind of force compatible with the isotropy of space: note that in an isotropic space without external forces

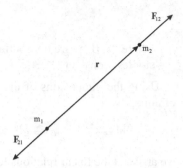

Fig. 4.2 Problem of two bodies. The vector **r** is the vector position of the particle of mass m_2 with respect to the position of another particle. In the case of an attractive force, the vectors $\mathbf{F}_{12}$ and $\mathbf{F}_{21}$ have opposite directions

the vector that connects the two particles has a unique preferred direction. Then, the type of force that we are considering is the most general one for the problem of two bodies.[4]

Newton's Second Law has the following form:

$$\mathbf{a}_1 = \frac{\mathbf{F}_{21}}{m_1}, \qquad \mathbf{a}_2 = \frac{\mathbf{F}_{12}}{m_2}.$$

One can rewrite these equations in terms of radius-vectors,

$$\ddot{\mathbf{r}}_1 = -\frac{\mathbf{F}}{m_1}, \qquad \ddot{\mathbf{r}}_2 = \frac{\mathbf{F}}{m_2}. \qquad (4.22)$$

The last system of equations is complicated because the force $\mathbf{F} = \mathbf{F}_{12}$ depends on both the position vectors, $\mathbf{r}_1$ and $\mathbf{r}_2$, and all six equations for the components of the position vectors are coupled. At the same time, the problem may be considerably simplified if we use other vectors, namely **r** and the radius-vector of the center of mass **R**, introduced in the previous section.

So, let us use the variables **r** and **R** instead of $\mathbf{r}_1$ and $\mathbf{r}_2$. The vector **R** is given by

$$\mathbf{R} = \frac{m_1\mathbf{r}_1 + m_2\mathbf{r}_2}{m_1 + m_2}. \qquad (4.23)$$

According to the law of conservation of momentum, we have $\dot{\mathbf{R}} = \mathbf{V} = $ constant. The second variable, **r**, is defined in (4.21). Equations (4.23) and (4.21) can be easily solved with respect to $\mathbf{r}_1$ and $\mathbf{r}_2$,

$$\mathbf{r}_2 = \mathbf{R} + \frac{m_1\mathbf{r}}{m_1 + m_2}, \qquad \mathbf{r}_1 = \mathbf{R} - \frac{m_2\mathbf{r}}{m_1 + m_2}. \qquad (4.24)$$

Now the problem becomes much simpler, since the dynamics of **R** has an obvious solution, given by

---

[4] The situation can be more complicated for the forces depending on velocities of the two particles, such as magnetic force. The consistent treatment of this case is possible only within relativistic electrodynamics, but the final output is in agreement with Eq. (4.21).

$$\mathbf{R}(t) = \mathbf{R}_0 + \mathbf{V}(t - t_0), \qquad \mathbf{R}_0 = \mathbf{R}(t_0).$$

Now we only need to explore the dynamics of $\mathbf{r}$. Using the Eq. (4.22) and the definition (4.21), we immediately obtain the result

$$\ddot{\mathbf{r}} = \ddot{\mathbf{r}}_2 - \ddot{\mathbf{r}}_1 = \frac{\mathbf{F}}{m_2} + \frac{\mathbf{F}}{m_1} = \frac{\mathbf{F}}{\mu},$$

$$\text{where} \quad \mu = \frac{m_1 m_2}{m_1 + m_2} \qquad (4.25)$$

is a quantity called "reduced mass" or "effective mass" of the system of two bodies.

One can note that, according to the formulas (4.25), the non-trivial part of the dynamical problem of the two bodies is reduced to the problem of an artificial particle of mass μ in an external central field given by the force $\mathbf{F}$,

$$\ddot{\mathbf{r}} = \frac{\mathbf{F}}{\mu}.$$

In this case, the expression "central field" means that the "particle" with mass μ interacts with the center of coordinates, placed at the point $\mathbf{r} = 0$. In fact, neither this particle, nor "the center" exist as a physical object, they only represent a useful tool for solving the motion of the system through the formulas (4.24).

It is important to remember that the force $\mathbf{F}$ depends only on the vector $\mathbf{r}$ and does not depend, in any way, on the radius-vector of the center of mass of the system $\mathbf{R}$. For this reason we are able to perform the change of variables leading to the factorization of the equations of motion, such that the equation for $\mathbf{r}$ does not depend on $\mathbf{R}$, and vice versa. Mathematically, our system of two interacting particles has been reduced to the dynamics of the two "artificial particles": one has radius-vector $\mathbf{r}$ and is moving in a central field of force $\mathbf{F}$ while another is free and has radius-vector $\mathbf{R}$.

For the interaction of a planet with a star (e.g., the Earth-Sun system), if we consider $m_2 \gg m_1$, then

$$\mu = \frac{m_1 m_2}{m_1 + m_2} \approx m_1.$$

Thus, it is a good approximation to take the Sun as a static body and assume that only the Earth is moving. We will study this problem with more details in Chap. 9.

Exercises

1. Check the formulas (4.24) and (4.25). Discuss the relationship between the change of variables assumed in these equations and the passage to the reference frame of the center of mass (C).

2. Consider the movement of the two point-like bodies of masses m and M, attached to a spring with elastic constant k, according to Fig. 4.3. The unstressed spring length is L, being different from this value at the initial time, and the initial velocities of the bodies are null. Find the solutions describing the motion for each body.

Hint. The problem is simple, since the initial velocities are null, hence there is no rotation. In this situation it is okay to consider the one-dimensional motion, so it is convenient to adopt the OX axis as the direction of the line connecting the bodies.

Answer:

$$x_1 = -\frac{M}{m+M}\left(L + A\cos\omega t\right) + X,$$

$$x_2 = \frac{m}{m+M}\left(L + A\cos\omega t\right) + X,$$

where X is the constant position of the center of mass and $\omega = \sqrt{k/\mu}$.

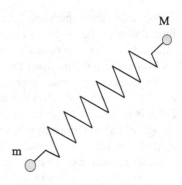

Fig. 4.3 Exercise 2. Two bodies connected by a spring

3. Consider the motion of three bodies of masses m_1, m_2, m_3 and radius-vectors $\mathbf{r}_1$, $\mathbf{r}_2$, $\mathbf{r}_3$. Suppose that

$$\mathbf{F}_{ij} = F_{ij}\frac{\mathbf{r}_i - \mathbf{r}_j}{|\mathbf{r}_i - \mathbf{r}_j|}$$

for $i, j = 1, 2, 3$ and $i \neq j$. Try to reduce the problem to the motion of two bodies using the reference frame of the center of mass (C). Show that a further reduction to the motion of a single particle is not possible.

Hint. The second reduction is not possible because the center of mass of the sub-system of the two bodies is moving, in the general case, with non-zero acceleration.

4. A system consists of two bodies with masses M and $2M$, moving with the respective velocities

$$\mathbf{v} = v_0\left(3\hat{\mathbf{i}} + 2\hat{\mathbf{j}}\right) \qquad \text{and} \qquad \mathbf{u} = v_0\left(\hat{\mathbf{i}} + 4\hat{\mathbf{k}}\right)$$

(a) Calculate the total momentum and the velocity of the center of mass.

(b) What is the distance traveled by the center of mass of the system during the time T?

(c) What is the kinetic energy of the system in the reference frame of the center of mass (C)?

Answers:

$$\text{(a)} \quad \mathbf{V}_c = \frac{v_0}{3}\left(5\mathbf{i} + 2\hat{\mathbf{j}} + 8\hat{\mathbf{k}}\right), \qquad \text{(c)} \quad K = 8Mv_0^2.$$

Observation. The reader who is not familiar with the concept of kinetic energy can look at the continuation of this problem in Chap. 6.

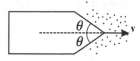

Fig. 4.4 Exercise 5

5. Consider the motion of a spacecraft through a cloud of static particles with the mass density ρ. The area of the front section of the ship is S and the angle between the front surface and the direction of motion is θ, as shown in Fig. 4.4. The particles collide elastically with the spacecraft. What is the force required to keep the speed v constant?

Solution. The angle of deviation of a particle of mass m is given by 2θ, so that the momentum transferred in the dispersion of each particle is $\Delta p = mv(1 - \cos 2\theta)$. During the time Δt the spacecraft collides with a number of particles given by $N = nSv\Delta t$. Here n is the number of particles per unit volume, such that $nm = \rho$. The force can be easily calculated as being

$$F = \frac{N\Delta p}{\Delta t} = 2\rho S v^2 \sin^2 \theta.$$

6. Find the position of the center of mass of a homogeneous flat circular body with a circular cavity, as shown in Fig. 4.5. Consider $R = 3m$ and $r = 1m$.

Solution. The most economic method of solving this problem is to consider the negative mass $-m$ with $m = M(r/R)^2$, in the place of the hole. The symmetry of the problem that indicates the center of mass must be on the straight line joining the centers of the two circles. Denoting the coordinate of the center of mass as L, we find

$$(M - m)L = 0 \cdot M - (R - r) \cdot m. \qquad \text{Then,} \quad L = -\frac{r^2}{R + r} = -0.25\,m.$$

Fig. 4.5 Exercise 6

7. Before the collision with the body B, the body A had the velocity **v** and the body B was at rest. After the collision, the body B had the velocity **u**, such that the angle between **v** and **u** is $\pi/3$ and also $|\mathbf{v}| = 2|\mathbf{u}|$. Determine the magnitude of the velocity of the body A after this collision, if the masses of the two bodies are equal.

Answer: $v_{after} = \sqrt{3}v/2$.

Chapter 5
Work of a Force and Conservative Forces

Abstract We are going to consider the work of a force and also introduce the notion of potential energy of a particle. Some important features related to these quantities will be used in order to classify different types of forces. Special attention will be given to some important forces, like gravitational force and air resistance, which is a dissipative force. We shall state that in a closed system the total work of all dissipative forces is negative.

5.1 Work of a Force

As we saw in the previous chapter, in a system of N particles, there are, in general, $6N$ integrals of motion (or conservation laws), some of them with important physical interpretation. For example, we know that the law of conservation of linear momentum is related to the homogeneity of space.

Let us recall that there are other important symmetries, such as isotropy of space and homogeneity of time. Therefore, besides the law of conservation of momentum, there should be other conservation laws, also with a clear physical interpretation. Our next subject is the conservation of energy, related to the homogeneity of time. This connection is demonstrated rigorously in Analytical Mechanics and here we approach the same problem from the more phenomenological side. The purpose of the present chapter is to introduce an important notion, which is called the work done by a force, that is useful to formulate the law of conservation of energy. Furthermore, we will perform a classification of forces that will be helpful in understanding this law.

Consider, as a first step, a particle in the external field, represented by the force $\mathbf{F}$. Let us suppose that this particle is moving from point 1 whose position is specified by the radius-vector $\mathbf{r}_1$ to the point 2, whose position is specified by $\mathbf{r}_2$, via a certain path, or trajectory, (L). This trajectory is illustrated in Fig. 5.1.

Let us now define the notion of *work of a force*, W_{12}, along the trajectory (L). For this purpose, we divide the path (L) into infinitesimal displacements $d\mathbf{r}$, such

I.L. Shapiro and G. de Berredo-Peixoto, *Lecture Notes on Newtonian Mechanics*,
Undergraduate Lecture Notes in Physics, DOI 10.1007/978-1-4614-7825-6_5,
© Springer Science+Business Media, LLC 2013

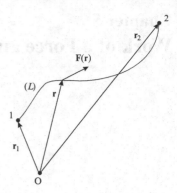

Fig. 5.1 Trajectory of a particle subjected to a force $\mathbf{F}(\mathbf{r})$

that the path length and the resulting displacement are given by the integral representations

$$L = \int_{(L)} dr \quad \text{and} \quad \mathbf{r}_2 - \mathbf{r}_1 = \int_{(L)} d\mathbf{r} \,. \tag{5.1}$$

The first expression is a line integral of the first type and the second one can be understood as

$$\int_{t_1}^{t_2} dt \left(\hat{\mathbf{i}} \frac{dx}{dt} + \hat{\mathbf{j}} \frac{dy}{dt} + \hat{\mathbf{k}} \frac{dz}{dt} \right) = \hat{\mathbf{i}} (x_2 - x_1) + \hat{\mathbf{j}} (y_2 - y_1) + \hat{\mathbf{k}} (z_2 - z_1),$$

where $\mathbf{r}(t)$ corresponds to the equation of the curve (L).

We can assume that the force vector $\mathbf{F}$ is a continuous function of coordinates. Then, in the infinitesimal range $d\mathbf{r}$, we can regard $\mathbf{F}$ as a constant vector. One can define the infinitesimal element of the work done by a force $\mathbf{F}$ (or infinitesimal work) dW as a scalar product

$$dW = \mathbf{F} \cdot d\mathbf{r} \tag{5.2}$$

and the total work done along the path (L) as the line integral of the second type,

$$W_{12} = \int_{(L)} dW = \int_{(L)} \mathbf{F} \cdot d\mathbf{r}. \tag{5.3}$$

If the force is generated by a field[1] we can also call W_{12} the work of the field. The expression for dW can be written in a more detailed form,

$$dW = F_1 dx + F_2 dy + F_3 dz = (F_1 v_x + F_2 v_y + F_3 v_z) \, dt \,,$$

---

[1] The more precise meaning of this will become clear later on, especially when we discuss potential forces.

where we have used the time as a parameter of the trajectory. It is evident that the same representation can be made with any other parameter along the trajectory. For example, in the case of the natural parameter, l, we obtain

$$dW = \mathbf{F} \cdot d\mathbf{r} = \mathbf{F} \cdot \frac{d\mathbf{r}}{dt} \frac{dt}{dl} \, dl = \mathbf{F} \cdot \frac{\mathbf{v}}{v} \, dl = \mathbf{F} \cdot \hat{\mathbf{v}} \, dl, \qquad (5.4)$$

where, as always, $dl = dr = |d\mathbf{r}|$ and

$$\hat{\mathbf{v}} = \hat{\mathbf{i}} \frac{dx}{dl} + \hat{\mathbf{j}} \frac{dy}{dl} + \hat{\mathbf{k}} \frac{dz}{dl} = \hat{\mathbf{i}} \cos\alpha + \hat{\mathbf{j}} \cos\beta + \hat{\mathbf{k}} \cos\gamma$$

and α, β, γ are the angles between the unit vector tangent to the trajectory, $\hat{\mathbf{v}}$, and the directions of the axes, vectors $\hat{\mathbf{i}}$, $\hat{\mathbf{j}}$ and $\hat{\mathbf{k}}$. Correspondingly, there are different representations for the work (5.3), e.g.,

$$W_{12} = \int_{t_1}^{t_2} \mathbf{F} \cdot \mathbf{v} \, dt = \int_0^L \mathbf{F} \cdot \hat{\mathbf{v}} \, dl$$

$$= \int_0^L dl \, (F_1 \cos\alpha + F_2 \cos\beta + F_3 \cos\gamma) \, . \qquad (5.5)$$

The last two expressions are ordinary integrals (Riemann definite integrals). Here we assume that the derivatives

$$\frac{dx}{dl} = \cos\alpha, \qquad \frac{dy}{dl} = \cos\beta, \qquad \frac{dz}{dl} = \cos\gamma$$

do exist and are continuous functions of a parameter l along the trajectory. Depending on the problem of interest, one or another form of an integral representation (5.5) can be the most useful one.

Example 1. Let us consider the work done by a homogeneous gravitational field, $\mathbf{g} = -g\hat{\mathbf{k}}$, when a mass m moves from the initial point 1 with the coordinates (x_1, y_1, z_1) to the final point 2 with the coordinates (x_2, y_2, z_2) (see Fig. 5.2).

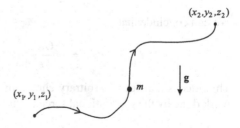

Fig. 5.2 The work done by the gravitational field along an arbitrary path between the two points (Example 1)

Solution. For an infinitesimal element of the work done by the gravitational force, $\mathbf{F} = m\mathbf{g}$, we have

$$dW = \mathbf{F}\cdot d\mathbf{r} = -mg\hat{\mathbf{k}}\cdot\left(\hat{\mathbf{i}}dx + \hat{\mathbf{j}}dy + \hat{\mathbf{k}}dz\right) = -mgdz,$$

i.e., the differential element of work depends only on the coordinate z (height). The work done by the force can be presented as

$$W_{12} = \int_{z_1}^{z_2} (-mg)dz = -mg(z_2 - z_1) = -mg\Delta z,$$

where Δz is the difference in heights of points 2 and 1. An important observation is that the work of the homogeneous gravitational field does not depend on the path but only on the initial and final points of the trajectory. Another way to state this is to say that the integral calculated over any closed path is zero,

$$\oint_c dW = \oint_c \mathbf{F}\cdot d\mathbf{r} = m\oint_c \mathbf{g}\cdot d\mathbf{r} = 0.$$

We will see that this property is valid for the gravitational and some other forces, however it is not valid in other cases, for example, for the force of friction.

Example 2. Let us consider the work done by the gravitational force generated by a point mass M, acting on a test particle of a small mass m. The force is given by Newton's gravitational law,

$$\mathbf{F} = -G\frac{Mm}{r^2}\hat{\mathbf{r}} = -G\frac{Mm}{r^3}\mathbf{r}, \qquad (5.6)$$

and the infinitesimal element of the work in the displacement $d\mathbf{r}$ is

$$dW = \mathbf{F}\cdot d\mathbf{r} = -\frac{GMm}{r^3}\mathbf{r}\cdot d\mathbf{r}.$$

We can simplify this relation by noting that the scalar product of any vector with itself is equal to the square of its absolute value, i.e., $\mathbf{r}\cdot\mathbf{r} = r^2$, where $r = |\mathbf{r}|$. Applying the differential operator, we obtain

$$\mathbf{r}\cdot d\mathbf{r} = \frac{1}{2}d(\mathbf{r}\cdot\mathbf{r}) = \frac{1}{2}d(r^2) = rdr. \qquad (5.7)$$

We can conclude that

$$dW = -\frac{GMm}{r^2}dr = GMmd\left(\frac{1}{r}\right).$$

This means that for an arbitrary choice of a path (L) between points 1 and 2, the work done by the gravitational force is given by

$$W_{12} = \int\limits_{(L)} dW = GMm \left(\frac{1}{r_2} - \frac{1}{r_1} \right).$$

In this case, just as in the previous example, the work does not depend on the path and is only defined by the initial and final points of the trajectory. The work along any closed path is zero, $\oint_c dW = 0$. Furthermore, the work done by the gravitational force is always zero for displacements of the test particle along a spherical surface with the center at the particle mass M. Only the change of the distance to the center implies zero work is done by the gravitational force. The situation is very similar to the case of homogeneous gravitational field (Example 1), where the work does not depend on the change of position of the particle along the coordinate axes OX and OY. For the case of the force (5.6), the work does not depend on the angular motions of the test particle. A more detailed consideration of the gravitational force can be found in the Appendix of this chapter.

Example 3. Let us consider the work done by a magnetic field[2] $\mathbf{B}$, so that $\mathbf{F} = e\mathbf{v} \times \mathbf{B}$, where $\mathbf{v}$ is the velocity of the particle which has the electric charge e. Remembering that $\mathbf{v} = d\mathbf{r}/dt$, we can write $F \perp d\mathbf{r}$. Therefore, $dW = \mathbf{F} \cdot d\mathbf{r} = 0$ for any $d\mathbf{r}$. This means that the magnetic force does not perform work. The result is valid for any field configuration.

It is clear that the effect of the magnetic force on the particle is not negligible, since the presence of the field $\mathbf{B}$ makes (in general) its trajectory curved. For example, if $\mathbf{B}$ is a constant and uniform field, and if $\mathbf{v} \perp \mathbf{B}$, the particle will travel along a circumference of radius R, defined by the equation

$$\frac{mv^2}{R} = evB, \quad \text{or} \quad R = \frac{mv}{eB}.$$

The biggest lesson we get from this example is that not all forces that do not perform work are irrelevant! The magnetic force is definitely important, with numerous technical applications, although it does not perform mechanical work.

5.2 Conservative Forces

All examples of forces considered until now have the property $\oint dW = 0$. This type of force is called *conservative*. We can distinguish two types of forces with this property, namely *potential force* and *solenoidal force*, as described below in what follows.

---

[2] The exact meaning of this expression is the work done by a magnetic force associated with the field $\mathbf{B}$.

5.2.1 Potential Force

A force **F** is called potential (one can say also that it is derived from a potential) when there is a scalar function, $U(\mathbf{r}) = U(x,y,z)$, such that the force is given by the gradient of this function,

$$\mathbf{F} = -\operatorname{grad} U(\mathbf{r}) = -\left(\hat{\mathbf{i}}\frac{\partial U}{\partial x} + \hat{\mathbf{j}}\frac{\partial U}{\partial y} + \hat{\mathbf{k}}\frac{\partial U}{\partial z}\right).$$

We can cite as an example of a potential force the gravitational force or Coulomb force in electrostatics. In the case of gravity, we have already found the expression for $U(\mathbf{r})$ in the Appendix of Chap. 3. Before returning to this case, we show that all potential forces are conservative.[3]

Let us consider again a curve (L) between point 1 and point 2 (see, for example, Fig. 5.1). Our goal is to show that the work of the potential force **F**, W_{12}, does not depend on the choice of the curve. We can write

$$W_{12} = \int\limits_{(L)} dW = -\int\limits_{(L)} \operatorname{grad} U(r) \cdot d\mathbf{r}. \tag{5.8}$$

In order to study this integral it is convenient to introduce a simple parameter λ along the curve, such that λ_1 corresponds to point 1 and λ_2 corresponds to point 2. We assume that each point of the curve corresponds to a certain value of λ, which increases with the motion from point 1 to point 2, i.e., $\lambda_1 \le \lambda \le \lambda_2$. The parameter λ can be time $(\lambda = t)$ or any other parameter, for example, the natural parameter l, but this choice is not really important for our consideration. We have

$$dW = \mathbf{F} \cdot d\mathbf{r} = -\operatorname{grad} U \cdot d\mathbf{r} \tag{5.9}$$

$$= -\left(\frac{\partial U}{\partial x}\frac{dx}{d\lambda} + \frac{\partial U}{\partial y}\frac{dy}{d\lambda} + \frac{\partial U}{\partial z}\frac{dz}{d\lambda}\right) d\lambda = -\frac{dU}{d\lambda} d\lambda = -dU(\lambda),$$

where in the last two equations we have considered $U = U(\mathbf{r}(\lambda))$ as a composite function of the parameter λ and used the chain rule for this function.

Consequently, for the integral (5.8), we obtain

$$W_{12} = -\int_{\lambda_1}^{\lambda_2} dU(\lambda) = U(\lambda_1) - U(\lambda_2). \tag{5.10}$$

It is clear that the work done by a potential force depends only on the values of the function U in the initial and final points of the path, but not on the details of the curve.

---

[3] Please note that the contrary is not always valid, i.e., not all the conservative forces are potential. An example is the *solenoidal force*, considered next.

In the case of a potential force, changing the direction of movement through the curve (L) means a change of sign of the work done by this force, as we can see in the Eq. (5.10). In addition, for a closed curve, the work of this force is always zero. The reason is as follows. Choose any two points (labeled 1 and 2) in a closed curve (see Fig. 5.3). Obviously, the integrals from point 1 to point 2 along the two paths are identical. But, in order to integrate along a closed curve, one has to calculate the integral from point 1 to point 2 and then go in the opposite direction, from point 2 to point 1, through another side of trajectory. Consequently, the result is null.

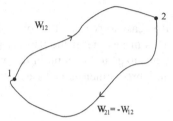

Fig. 5.3 The work done by a potential force along a closed path is always zero

As we can see, the case when the force is potential, $\mathbf{F} = -\operatorname{grad} U(r)$, is quite special. The function $U(r)$ is called the *potential energy function* related to the force $\mathbf{F}$, or *potential energy* or simply *potential* of the particle. For the Coulomb and gravitational fields one can make consideration in terms of the field generated by the masses or by external charges. When a test particle of mass m or charge e is placed at a point in space where there is a field, there is, typically, a potential energy.

An important aspect is that only the gradient of the potential energy, $\operatorname{grad} U(\mathbf{r})$, and not $U(\mathbf{r})$ itself, has a physical meaning. Therefore, the choice of zero energy is entirely arbitrary and should be chosen in a practical and convenient way. For example, one usually defines the gravitational potential energy associated with the gravitational force generated by the mass M as

$$U(\mathbf{r}) = -\frac{GMm}{r},$$

so that $U(r = \infty) = 0$. But it is perfectly possible to redefine $\tilde{U}(\mathbf{r}) = U(\mathbf{r}) + C$, where C is an arbitrary constant. Independently of the value of C, the force $\mathbf{F} = -\operatorname{grad} U = -\operatorname{grad} \tilde{U}$ is the same. One particularly simple example is as follows. As we have discussed before, $U(z) = mgz$ is the potential energy for the homogeneous gravitational field. Here the constant C is chosen such that $U = 0$ on the level $z = 0$. At the same time, one can choose $U(z) = mg(z - z_0)$ with an arbitrary z_0 and the gravitational force does not change.

Example 4. Consider the potential $U(x)$ (one-dimensional) shown in Fig. 5.4.

We know that the force on the particle subject to $U(x)$ is given by the expression

$$F(x) = -\frac{dU}{dx}.$$

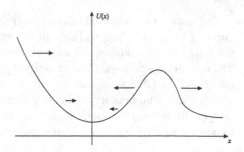

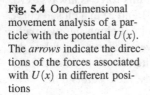

Fig. 5.4 One-dimensional
movement analysis of a particle with the potential $U(x)$.
The *arrows* indicate the directions of the forces associated
with $U(x)$ in different positions

It is easy to see that the force always acts in the direction of decreasing potential.
This means that if the particle is at rest near the minimum of potential, it always
tends to go to the point of the minimum. Qualitatively, the same occurs in the two-
and three-dimensional cases.

5.2.2 Solenoidal Force

A force is solenoidal when its work is identically zero, $dW \equiv 0$, regardless of
the displacement vector $d\mathbf{r}$. An example is the magnetic force. The solenoidal
force is always orthogonal to the velocity or, in other words, to the infinitesimal
displacement of the particle.

The following observation is in order. It is perfectly possible to have a motion
derived from a potential force that does not perform work, even if this force is essential to the movement. For example, consider the circular motion of a satellite
around a planet. The gravitational force is the centripetal force that ensures the circular motion. But in all instances the force is perpendicular to velocity, so that the
work done by the gravitational force is zero. This does not make gravity a solenoidal
force, of course. The fact that the force is solenoidal or potential is a fundamental
property of the force itself and does not depend on the particular motion of the
particle in a given problem.

5.3 Dissipative Forces

We now consider examples of other forces, which are not described by fields. We
begin with a particular example.

Example 5. Consider a body of mass m moving on a horizontal plane, with the
friction force $\mathbf{F}_{at}$ between the surfaces. The direction of $\mathbf{F}_{at}$ is always opposite to
the infinitesimal displacement, $d\mathbf{r}$, and its modulus is given by μN, where $N = mg$
is the normal force that the plane exerts on the body and μ is called the coefficient
of friction (see Fig. 5.5).

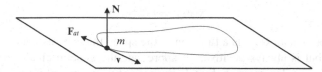

Fig. 5.5 Motion of a body on a surface with friction (Example 5). The work done by the frictional force along a closed path is different from zero, in contrast with the work done by conservative forces

The direction of the frictional force depends on the direction of motion of the body. We can write this force as

$$\mathbf{F}_{at} = -\mu mg \frac{d\mathbf{r}}{dr} = -\mu mg \,\hat{\mathbf{v}},$$

where $dr = |d\mathbf{r}|$. Thus, the differential element of the work is given by

$$dW = -\mathbf{F}_{at} \cdot d\mathbf{r} = -\mu mg \frac{d\mathbf{r} \cdot d\mathbf{r}}{dr} = -\mu mg \frac{dr^2}{dr} = -\mu mg dr, \quad (5.11)$$

that is, it is proportional to the modulus of the infinitesimal displacement of the particle. Integrating over the path of the motion, we obtain, using Eq. (5.1),

$$W_{12} = -\int_{(L)} \mu mg \, dr = -\mu mg L.$$

One can see that for any trajectory on the plane, the work done by the frictional force is proportional to the length of the curve and is always negative.

Example 6. As we already know the force of air resistance (or of some other fluid) for small speeds is proportional to the speed of the movement of the body, $\mathbf{F}_r = -\alpha \mathbf{v}$, where $\alpha = const$ and $\alpha > 0$. In this case, noting that $d\mathbf{r} = \mathbf{v} dt$, one can write

$$dW = \mathbf{F} \cdot d\mathbf{r} = -\alpha \mathbf{v} \cdot d\mathbf{r} = -\alpha \mathbf{v} \cdot \mathbf{v} dt = -\alpha v^2 dt.$$

Then, for a movement along any curve, (L), we find

$$W(L) = -\alpha \int_{t_1}^{t_2} v^2(t) dt < 0.$$

Obviously, in the last two examples, the work done by a force along a closed trajectory is not zero, i.e., the forces of surface friction and air resistance are not conservative.

Now we can formulate a general statement about the work done by *dissipative forces*. In general, these forces may be represented by the formula

$$\mathbf{F} = -k(v)\mathbf{v} = -k(v)v\hat{\mathbf{v}}, \tag{5.12}$$

where the coefficient $k(v)$ is a function of the speed that can also depend on other parameters, but is always positive. To appreciate the general character of (5.12), consider the examples treated before. In the case of the frictional force between two surfaces, the magnitude of the dissipative force does not depend on the modulus of velocity, being proportional to μN. In this case, $k(v) = \mu N/v$.

The last formulas are valid when the body is moving. When it is at rest, the unit vector $\hat{\mathbf{v}}$ is not defined and the presence of a friction force does not imply the production of mechanical work. This does not necessarily mean that the friction force vanishes. For example, the friction force can be nonzero, although its direction and orientation depend on other forces. The important thing is that when the body is at rest, the forces acting on this body, including dissipative forces, are not doing any work.

A very important feature of dissipative forces can be formulated in the form of the following Theorem:

Theorem 1. *For a closed system (free of external forces), the total work of all dissipative forces is always negative.*

Proof. In order to show the validity of this assertion, we recall that, according to Newton's Third Law, the forces in a closed system can be always separated into pairs. Therefore, we can consider a system with only two bodies, without harming the generality of the demonstration. Suppose that these two bodies are moving with velocities $\mathbf{v}_1$ and $\mathbf{v}_2$, and that there is a dissipative force between them. In this case the element of the differential work of the dissipative force is given by the expression

$$dW_{\text{diss}} = (\mathbf{F}_1 \cdot \mathbf{v}_1 + \mathbf{F}_2 \cdot \mathbf{v}_2)\, dt, \tag{5.13}$$

where, due to Newton's Third Law, $\mathbf{F}_2 = -\mathbf{F}_1$. Because the forces are dissipative, their magnitudes are dependent on the corresponding velocities. But since there are only two bodies, the unique relevant velocity is the relative one, $\mathbf{v} = \mathbf{v}_1 - \mathbf{v}_2$. The dependence on any other velocity would mean a violation of Galileo's principle, i.e., the existence of some preferred inertial reference system. In other words, the force $\mathbf{F}_1$ can only depend on the difference $\mathbf{v}_1 - \mathbf{v}_2$, because this force is caused by the interaction of the first body with the second body and can only depend on their relative speed. Then the forces acting on the bodies are written in the form

$$\mathbf{F}_1 = -\mathbf{F}_2 = -k(v)\mathbf{v}.$$

In this case, Eq. (5.13) becomes

$$\begin{aligned}
dW_{\text{diss}} &= -k(v)\left[(\mathbf{v}_1 - \mathbf{v}_2) \cdot \mathbf{v}_1 - (\mathbf{v}_1 - \mathbf{v}_2) \cdot \mathbf{v}_2\right] dt \\
&= -k(v)\mathbf{v} \cdot (\mathbf{v}_1 - \mathbf{v}_2)\, dt = -k(v)\mathbf{v} \cdot \mathbf{v}\, dt = -k(v) \cdot v^2 dt < 0. \tag{5.14}
\end{aligned}$$

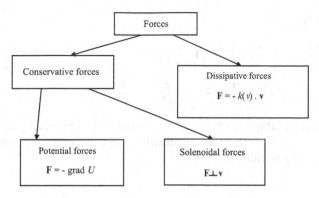

Fig. 5.6 General classification of forces

The next observation is that the result (5.14) does not depend on the choice of the inertial reference frame, i.e., the laboratory one. This is due to the fact that this relation include only the relative speeds of the bodies.

At last, we consider a closed system of many bodies. Similar to the previous consideration, the forces that act in this system can be divided into pairs and the total work accomplished by the forces of the system is the sum of the work performed by each pair of forces. Using the result (5.14) for each pair, we conclude that the same inequality, $dW_{diss} < 0$, remains valid for the many-body system. $\square$

Finally, we can classify all the forces according to the scheme illustrated in Fig. 5.6.

Appendix

We will show that the gravitational force,

$$\mathbf{F} = -\frac{GMm}{r^3}\mathbf{r} = -\frac{GMm}{r^2}\hat{\mathbf{r}}, \tag{5.15}$$

admits the potential

$$U(r) = -\frac{GMm}{r}.$$

For this end we will use Cartesian coordinates, recalling that $r = \sqrt{x^2+y^2+z^2}$. To calculate the gradient of $U(r)$, we take the partial derivatives,

$$\frac{\partial r}{\partial x} = \frac{2x}{2r} = \frac{x}{r}, \qquad \frac{\partial r}{\partial y} = \frac{2y}{2r} = \frac{y}{r}, \qquad \frac{\partial r}{\partial z} = \frac{2z}{2r} = \frac{z}{r}.$$

Thus, we can write

$$
\nabla\left(\frac{1}{r}\right) = \left(\hat{\mathbf{i}}\frac{\partial}{\partial x} + \hat{\mathbf{j}}\frac{\partial}{\partial y} + \hat{\mathbf{k}}\frac{\partial}{\partial z}\right)\frac{1}{r} = -\frac{1}{r^2}\left(\hat{\mathbf{i}}\frac{\partial r}{\partial x} + \hat{\mathbf{j}}\frac{\partial r}{\partial y} + \hat{\mathbf{k}}\frac{\partial r}{\partial z}\right)
$$

$$
= -\frac{\hat{\mathbf{i}}x + \hat{\mathbf{j}}y + \hat{\mathbf{k}}z}{r^3} = -\frac{\mathbf{r}}{r^3} = -\frac{\hat{\mathbf{r}}}{r^2}.
$$

Correspondingly, multiplying by $-GMm$, we obtain (5.15), exactly as one should expect. It is easy to see that this calculation is directly related to the procedure considered in the Appendix of Chap. 3.

Exercises

1. Consider a central force,

$$
\mathbf{F} = f(r)\cdot\mathbf{r},
$$

where $f(r)$ is a continuous function of the modulus $r = |\mathbf{r}|$. Show that this force is conservative and find an integral representation for $U(r)$ such that $\mathbf{F} = -\nabla U$.

Answer.

$$
U(r) = -\int rf(r)dr = -\int_{r_0}^{r} r'f(r')dr'.
$$

2. In the previous problem, consider the following special cases:

(a) $f(r) = r^{-n}$;

(b) $f(r) = \frac{1}{r}e^{-\alpha r}$, where $\alpha < 0$.

For these two cases, calculate $U(r)$. Look for a physical interpretation for the cases $n = 2$ and $n = 3$ in item (a). Sketch the graph of $U(r)$ for both values of n.

Answers:

$$
(a)\qquad U(r) = \frac{r^{2-n}}{n-2} + \text{const.}\quad\text{for}\quad n \neq 2;
$$

$$
U(r) = -\ln\left(\frac{r}{r_0}\right)\quad\text{for}\quad n = 2.
$$

$$
(b)\qquad U(r) = \frac{1}{\alpha}e^{-\alpha r} + const.
$$

For $n = -2$, there is no position of stable equilibrium and the movement is not limited in the radial coordinate.

3. For the potential $U(\mathbf{r}) = ax^n + by^n + cz^n$, where $n = -1, 0, 1, 2, 3$, calculate the associated force. Formulate criteria for the parameters a, b, c, such that

(a) The field is central.
(b) The force is attractive to the center.
(c) The test particle can reach the center $\mathbf{r} = 0$.

Answer:
For $n = 0$ the force is zero.
(a) For the field to be central, it is necessary that $a = b = c$ and also $n = 2$.
(b) $n = 2$, also $a, b, c > 0$.

4. A particle of mass m is moving according to the equations

$$\mathbf{r}(t) = v_0 \left(t\hat{\mathbf{i}} + 3t\hat{\mathbf{j}} - \frac{5t^2}{t_0} \hat{\mathbf{k}} \right).$$

Which components of the momentum of the particle do not depend on time? Find the components of force $\mathbf{F}$ acting on the particle.

Answers:
The components p_x and p_y are constants; $F_z = -10v_0 m/t_0$.

5. For the mechanical system from the previous problem, determine the potential $U(\mathbf{r})$, such that the force should be represented as $\mathbf{F} = -\,\text{grad}\,U(\mathbf{r})$. Calculate the work done by the force $\mathbf{F}$ between the instants t_1 and t_2.

Answers:

$$U = \frac{10v_0 m}{t_0} z + const ; \qquad W_{12} = \frac{50 m v_0^2}{t_0^2} \left(t_2^2 - t_1^2 \right).$$

6. A particle was displaced from the point whose radius-vector is given by $\mathbf{r}_1 = 3r_0\hat{\mathbf{i}} - r_0\hat{\mathbf{j}}$ to the point with the radius vector, $\mathbf{r}_2 = r_0\hat{\mathbf{j}} + 2r_0\hat{\mathbf{k}}$.
On the way this particle was subject to the force $\mathbf{F} = F_1\hat{\mathbf{i}} + F_2\hat{\mathbf{j}} + F_3\hat{\mathbf{k}}$, where all the components F_1, F_2 and F_3 are constants. Calculate the work done by the force $\mathbf{F}$ in this process. Does the result depend on the choice of path?

Answer:
The work $W_{12} = r_0 \left(-3F_1 + 2F_2 + 2F_3 \right)$ is independent of the choice of a path.

7. A particle of mass M is moving according to the equations

$$x = R\cos \omega t , \qquad y = R\sin \omega t , \qquad z = v_0 t .$$

Find the total force acting on the particle and classify this force, indicating whether it is conservative or not. If so, is it possible to determine, for this case, if the force is solenoidal or potential?

Answer:
One can directly check that the force $\mathbf{F} = -m\omega^2 (x\hat{\mathbf{i}} + y\hat{\mathbf{j}})$ is perpendicular to the velocity of the particle. This means that the work done by the force is zero and hence it is conservative. At the same time, one can not define whether this force has potential or solenoidal nature, because the same acceleration can be produced by magnetic force (solenoidal) but also by the electric force of attraction due to a linear uniformly charged wire situated on the axis $x = y = 0$.

Chapter 6
Conservation of Energy

Abstract The conservation of energy is one of the most important principles of physics. In this chapter, our goal is to discuss this conservation law in the framework of Classical Mechanics where it can be obtained from Newton's Laws. We can distinguish, in principle, two different cases, in which the law of conservation of energy applies, one of them includes closed systems, and the other, systems which are subject to the action of conservative external forces. We shall begin our consideration with the study of a relatively simple situation in which there is only one particle subject to an external field of a potential force. Then we will study more complicated case of a closed system of various particles. It is worth warning that the approach taken in this chapter is not the best possible, from mathematical and physical points of view. A more complete and formal consideration can be performed in Analytical Mechanics, where it is possible to relate the conservation laws with fundamental symmetries of space and time. In particular, the law of conservation of energy is related to the homogeneity of the time, but at the level of the present book, it is difficult to recognize this connection.

6.1 Particle in a Potential Force Field

As a first step, consider a particle subject to external forces, all conservative by assumption. In this case, the equation of motion is the realization of Newton's second law. Solenoidal forces, if they exist, are not relevant to our consideration, because they do not produce any effects on the conservation of energy. Therefore, without loss of generality, one can write

$$\mathbf{F} = m\ddot{\mathbf{r}} = -\operatorname{grad} U(\mathbf{r}).\tag{6.1}$$

Let us consider the movement between the instants t_1 and t_2, such that $\mathbf{r}(t_1) = \mathbf{r}_1$ and $\mathbf{r}(t_2) = \mathbf{r}_2$. Assume that the curve (L) is a trajectory of the particle between

I.L. Shapiro and G. de Berredo-Peixoto, *Lecture Notes on Newtonian Mechanics*,
Undergraduate Lecture Notes in Physics, DOI 10.1007/978-1-4614-7825-6_6,
© Springer Science+Business Media, LLC 2013

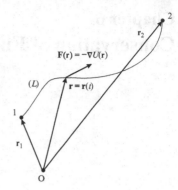

Fig. 6.1 Trajectory of a particle under the influence of conservative forces

the points $\mathbf{r}_1$ and $\mathbf{r}_2$, described by the vector $\mathbf{r}(t)$, which is a solution of (6.1). The situation is illustrated in Fig. 6.1.

We already know that the integration of the right hand side part of (6.1) along the trajectory (L) is given by

$$-\int_{(L)} \mathbf{F} \cdot d\mathbf{r} = -W_{12} = \int_{(L)} \operatorname{grad} U \cdot d\mathbf{r} = U(2) - U(1), \qquad (6.2)$$

where we use the compact notation $U(1) = U(\mathbf{r}_1)$ and $U(2) = U(\mathbf{r}_2)$. Obviously, it is very interesting to see what happens with the left side of the Eq. (6.1) under the same integration. As we shall see in a moment, the fact that this integration results in a difference between a function at points 1 and 2 corresponds to a new integral of motion, i.e., to a new conservation law. Using Newton's Second Law, we obtain

$$W_{12} = \int_{(L)} \mathbf{F} \cdot d\mathbf{r} = \int_{(L)} m\ddot{\mathbf{r}} \cdot d\mathbf{r} = \int_{(L)} m\frac{d\mathbf{v}}{dt} \cdot d\mathbf{r} = \int_{(L)} m d\mathbf{v} \cdot \frac{d\mathbf{r}}{dt} \qquad (6.3)$$

$$= \int_{(L)} m\mathbf{v} \cdot d\mathbf{v} = \int_{(L)} d\left(\frac{m\mathbf{v} \cdot \mathbf{v}}{2}\right) = \int_{(L)} d\left(\frac{mv^2}{2}\right) = K(2) - K(1),$$

where we have used the relation $\mathbf{v} \cdot \mathbf{v} = v^2$. The new function K, defined by

$$K = \frac{mv^2}{2}, \qquad (6.4)$$

is called *kinetic energy* of the particle. The formula (6.3) is known as the *theorem of kinetic energy*: the total work done on a particle equals the change of its kinetic energy. It is easy to note that the kinetic energy K is very different from the potential energy, U. For example, the expression for K only shows the dependence on the velocity of the particle, as opposed to the formula for U, which depends only on the position. Nevertheless, the differences between the two quantities are numerically equal. Summing up the Eqs. (6.2) and (6.3), we can observe that the result is zero.

What we get here is that for the motion of a particle in an external potential field, governed by Newton's Second Law (6.1), for any moment t_1 and t_2, the following relation holds $K(2) - K(1) = U(1) - U(2)$ and, therefore,

$$K(1) + U(1) = K(2) + U(2).$$

This means that the sum of kinetic energy with the potential energy is constant:

$$E = \frac{mv^2}{2} + U(\mathbf{r}) = \text{const.}, \qquad (6.5)$$

This quantity is called mechanical energy or simply energy of the particle. An important property of the kinetic energy is that it is defined to be an additive quantity. The kinetic energy of a system of particles is the sum of the kinetic energies of its parts, e.g., sum of the kinetic energies of the individual particles.

The first observation is that we can also include solenoidal forces. The main point is that these forces have the property $dW = 0$ and therefore neither K nor U changes during the movement. So, the law (6.5) remains valid.

The way in which we obtained the law of energy conservation is very particular, but it can be generalized to many other situations. After all, energy conservation represents perhaps the most fundamental law of physics, and is valid throughout the physical and consequently in all other sciences such as chemistry and biology, where some other laws of physics do not apply. In relativistic theory, the law of conservation of energy is formulated in conjunction with the conservation of momentum (one can say that both are mixed in a single law of conservation).

In another area of Physics, energy conservation is called the First Law of Thermodynamics. Finally, in Cosmology, one can study the history of the Universe from the mysterious moment of its creation. Modern Cosmology is an extremely complicated science, which is based on the approaches and results of many other areas of physics. The whole Universe as an object of study is really complicated. However, energy conservation still holds and helps to develop this area of Physics, in many cases bringing results which are experimentally confirmed. The result (6.5) we get here is very simple and the path to the applications mentioned above is very long. In the next sections we will try to generalize the law of conservation of energy in Classical Mechanics.

Exercises

1. Verify the relation (5.7) explicitly using Cartesian and spherical coordinates. Show that a similar formula is valid for velocities.

2. Consider the one-dimensional movement of a point-like body of mass m in the external potential $U(x)$, such that the force is given by

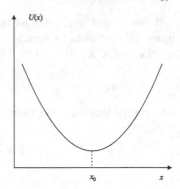

Fig. 6.2 Point particle in the
one-dimensional potential
$U(x)$ (Exercise 2)

$$F = -\frac{dU}{dx} = -U'. \tag{6.6}$$

Note that in the one-dimensional case, it is not necessary to use the arrow notation
for vectors because they have a single component. Perform the following tasks:
(a) Write the law of conservation of energy, relating x and $v = \dot{x}$, in the form
$E = K + U = constant$.
(b) Using the chain rule, take the time derivative of the energy and show that the law
$E = K + U = constant$ produces Newton's Second Law, $m\ddot{x} = F$.
(c) Consider the case where the potential satisfies the condition

$$\lim_{|x| \to \infty} U(x) = +\infty$$

and has a unique point of minimum, x_0 (see Fig. 6.2). In this case, $U' > 0$ for $x > x_0$
and $U' < 0$ for $x < x_0$. Write some examples of expressions for $U(x)$ which
would have these two properties. Discuss qualitatively the motion of a particle in
this potential using both the conservation of energy, and Newton's Second Law.
(d) Consider the motion of a particle with energy $E = constant$ and write the
equations for points x_1 and x_2, where the kinetic energy is zero, i.e., where the
particle instantly stops its motion. These points are called turning points, or return
points.
(e) Write an integral representation for the period of oscillations, T, between the
points of return in the part (d).

Answers :(d,e). $U(x_1) = U(x_2) = E$, $T = \sqrt{2m} \int_{x_1}^{x_2} \frac{dx}{\sqrt{E - U(x)}}. \tag{6.7}$

Hint. In case of (e) one can obtain the result immediately, after using the energy
conservation law

$$\frac{m\dot{x}^2}{2} + U(x) = E = const$$

and integrating between the return points. It is interesting to note that the integral in (6.7) is the so-called improper integral, because the corresponding integrand $(E - U)^{-1/2}$ becomes infinite in the vicinity of both return points. The convergence of this integral is related to some physical conditions. We leave it to the reader to explore this issue as an additional (relatively complicated) exercise.

6.2 Conservation of Energy in Closed Systems

The next step is to consider a closed system of many particles. We consider first the system of two point-like bodies, 1 and 2, with masses m_1 and m_2, interacting with each other through a conservative force satisfying Newton's third law, $\mathbf{F}_{12} = -\mathbf{F}_{21}$. Since the system is closed, any change in the kinetic energy of the body 1 can be only due to the force $\mathbf{F}_{21}$, and vice versa – the kinetic energy of the body 2 changes only due to the force $\mathbf{F}_{12}$.

Consider the movement between two instants of time t_a and t_b, in which the positions of the particles 1 and 2 are specified by $\mathbf{r}_a^{(1,2)}$ and $\mathbf{r}_b^{(1,2)}$ correspondingly. The change of kinetic energy of the body 1 is equal to the work of the force $\mathbf{F}_{21} = m_1 \ddot{\mathbf{r}}_1$,

$$K_b^{(1)} - K_a^{(1)} = W_1 = \int_{(ab)} \mathbf{F}_{21} \cdot d\mathbf{r}_1 = \int_{(ab)} m_1 \ddot{\mathbf{r}}_1 \cdot d\mathbf{r}_1 \,.$$

For the body 2, similarly, one can write

$$K_b^{(2)} - K_a^{(2)} = W_2 = \int_{(ab)} \mathbf{F}_{12} \cdot d\mathbf{r}_2 = \int_{(ab)} m_2 \ddot{\mathbf{r}}_2 \cdot d\mathbf{r}_2 \,.$$

Summing up the last two equation, one can easily see that the change of the total kinetic energy of the system,

$$K = K^{(1)} + K^{(2)} = \frac{m_1 v_1^2}{2} + \frac{m_2 v_2^2}{2} \,,$$

is equal to the work performed by all the forces of the system, $W = W_1 + W_2$. So,

$$K_b - K_a = W_{ab} \,, \tag{6.8}$$

where K_a is the total kinetic energy at the time t_a and K_b is the total kinetic energy at the time t_b. W_{ab} is the work performed by both forces between these two instants. Moreover, the same work also causes change in the potential energy of the system,

$$W_{ab} = W_{ab}^{(1)} + W_{ab}^{(2)} = U_a - U_b \,.$$

This relation is true in the case of potential forces, and the demonstration of this fact is similar to the previous case of a particle in the field of potential force, so we leave it as an exercise.

Finally, we can note that

$$E_a = U_a + K_a = U_b + K_b = E_b,$$

i.e., the total mechanical energy of the system is an integral of motion.

The consideration presented above can be generalized to a system of N particles. Let us remember that all forces come in pairs, such that $\mathbf{F}_{ij} = -\mathbf{F}_{ji}$, for $i \neq j$. The same treatment which we presented above can be applied to each pair of forces and the results will be the same. Adding the contributions of all pairs of forces one arrives at the conclusion that $K_b - K_a = W_{ab}$ and also $U_a - U_b = W_{ab}$, where W_{ab} is the total work done by all the internal forces of the system.

Furthermore, we can include, using the results of the previous section, the potential and solenoidal external forces (the last do not perform mechanical work and hence have no essential role here). The total energy, $E = K + U$, will remain an integral of motion.

Finally, we note that the potential energy of a system of N particles can be written as

$$U(\mathbf{r}_1, \ldots, \mathbf{r}_N) = U_{int} + U_{ext}, \tag{6.9}$$

where $U_{int} = U_{int}(\mathbf{r}_1, \ldots, \mathbf{r}_N)$ is potential energy of interaction between the particles of the system and $U_{ext} = U_{ext}(\mathbf{r}_1, \ldots, \mathbf{r}_N)$ is potential energy of interaction with external fields. The latter ones are generated by the bodies (or charges etc) which do not belong to the system.

When dissipative forces are present, the mechanical energy, $K + U$, is not an integral of motion, so the law of conservation for mechanical energy is not valid. Within a broader perspective, energy is indeed preserved, but it is not purely mechanical energy anymore. So, in order to understand this phenomenon, one must take into account other forms of energy. For example, any force of friction (including surface friction, air resistance, etc.) transfers mechanical energy into heat which becomes an internal energy of the bodies. However, this energy is also the sum of kinetic and potential energies of the molecules and atoms, of which all bodies are composed. Taking all these factors into account, we can establish the universality of energy conservation.

Exercises

1. Show that $W_{ab} = U_a - U_b$ in the case of a closed system of many particles interacting by means of potential forces.

Hint. Try to use the result of the problem of two bodies and consider the motion in an external field created by other particles.

2. Discuss the following statements, indicating whether each of them is right or wrong. Consider the reference frame of the laboratory. Justify your answers.

(a) The work of a dissipative force for a body in motion can not be null.
(b) Only conservative forces perform work.
(c) The work done by a conservative force acting on a body is always equal to the variation of the kinetic energy of the body.
(d) The work done by a conservative force is always equal to the decrease in the potential energy associated to this force.

Answers: *(a)* Right; *(b)* Wrong; *(c)* Wrong; *(d)* Right.

Observation: These answers correspond to an inertial reference frame and may be different in an arbitrary reference frame. For example, in the proper frame of the particle which suffers the action of a dissipative force, the work of this force is zero.

3. A heavy block of mass M moves on a horizontal plane, under the action of the force **F** exerted by a rope. The force makes an angle φ relative to the plane. The friction coefficient between the block and the plane is given by μ.

a) Calculate the work of the force **F**, of the frictional force and also the gravity force, when the block travels a distance L along the plane.
b) For which value of φ an acceleration of the block achieves its maximum?
c) Does this value correspond to the minimum work of frictional force? Try to answer this question also qualitatively.

Answers:

(a) $W_F = FL\cos\varphi$; $\quad W_{friction} = -\mu(Mg - F\sin\varphi)L$; $\quad W_{grav} = 0$.
(b) The acceleration is given by

$$a = \frac{F}{M}\sqrt{1+\mu^2}\,\cos(\varphi - \theta) - \mu g,$$

where $\theta = \arctan\mu$. It is maximal when $\varphi = \theta$.
(c) No.

4. A block of mass M moves along a horizontal plane without friction. The initial velocity of the block is v_0. At a moment, the block enters a rough region of length L, where the coefficient of friction is μ. What should be the minimum initial velocity of the block so that it can cross this region?

Answer : $v_0 \geq \sqrt{2\mu gL}$.

5. A particle of mass m has horizontal velocity v when it collides with a block of mass M at rest, which has an inclined plane surface. An elastic collision occurs on the inclined plane such that the particle shortly after the impact has a vertical velocity relative to the block (note that its velocity is not vertical in relation to the ground). Calculate the height reached by the particle in relation to the horizontal line of its original trajectory. Discard the friction between the block and the floor.

Observation. Elastic collision means the mechanical energy is preserved. The reader will meet more details on this issue in the next sections.

Answer : $\quad h = \dfrac{v^2}{2g}\dfrac{M}{M+m}.$

6. A container of mass M is on a scale and receives a fluid that falls in it from the height h at a constant rate of μ grams per second. Initially the fluid is at rest. Suppose that the fluid is at rest again after it reaches the container (disregard turbulence), which has dimensions much smaller than h. Determine the reading of the scale when the mass m of the fluid has been drained into container.

Answer : $\quad M + m + \mu\sqrt{\dfrac{2h}{g}}.$

6.3 Kinetic Energy in Different Reference Frames

Before considering some examples of application of the energy conservation law, let us consider the transformation of the kinetic energy under a change of inertial reference frame. In particular, we focus our attention on the transformation between center of mass (C) and the laboratory (L) reference frames.

To start, consider a single particle. It is easy to see that the kinetic energy of the particle depends on the choice of inertial reference frame. When we pass from a reference system $\mathcal{K}$ to another one $\mathcal{K}'$ (moving with constant velocity $\mathbf{V}$ in relation to $\mathcal{K}$), the kinetic energy is transformed as follows:

$$ K' = \frac{mv'^2}{2} = \frac{m}{2}\mathbf{v}'\cdot\mathbf{v}' = \frac{m}{2}\left(\mathbf{v}-\mathbf{V}\right)\cdot\left(\mathbf{v}-\mathbf{V}\right). $$

Obviously, the value of the kinetic energy may be different in different frames. Incidentally, the same applies to the potential energy, as we have discussed above. The value of this energy depends on the choice of zero level. For example, consider a pen of mass $m = 100\,\text{g}$ placed on a table with height $h = 1\,\text{m}$ relative to the ground. The potential energy of the pen relative to the ground is $U = mgh = 0.98\,\text{J}$, but its potential energy in relation to the surface of the table is zero. If the pen falls down from the table, neither the increase of its speed, nor the gravitational force, or the acceleration during the fall, depend on this arbitrariness. These examples show that the energies K, U and E have an ambiguity related to the choice of the reference frame. The origin of this ambiguity lies in the fact that the energies can not be observed directly. What can be observed are forces, acceleration, speed changes or work of the forces. The latter represent, as it was already noted, differences in the values of the energies at the beginning and at the end of the process. Therefore, the physical observables are free of ambiguities related to the choice of reference system. Later on, we will simply choose the most useful framework to solve problems. To do this, it would be interesting to see how the kinetic energy of a system of many particles is changed under transformation of reference frame.

We consider that at the frame of reference (L) a particle of mass m_i ($i = 1, 2, \ldots, N$) has velocity $\mathbf{v}_i$. At the same reference frame (L), the center of mass of the system has velocity

$$\mathbf{V} = \frac{m_1\mathbf{v}_1 + m_2\mathbf{v}_2 + \ldots + m_N\mathbf{v}_N}{m_1 + m_2 + \ldots + m_N}.$$

Then the velocity of the particle of mass m_i in the system (C) is given by

$$\mathbf{v}'_i = \mathbf{v}_i - \mathbf{V}. \tag{6.10}$$

For the kinetic energies, we have the relations

$$K = \sum_{i=1}^{N} \frac{m_i v_i^2}{2} \quad \text{and} \quad K_c = \sum_{i=1}^{N} \frac{m_i v_i'^2}{2}, \tag{6.11}$$

where the index c indicates the reference (C).

Taking into account (6.10), we can rewrite the kinetic energy K in the frame (L), simply by noting that

$$v_i^2 = \mathbf{v}_i \cdot \mathbf{v}_i = (\mathbf{v}'_i + \mathbf{V}) \cdot (\mathbf{v}'_i + \mathbf{V}) = v_i'^2 + 2\mathbf{V} \cdot \mathbf{v}'_i + V^2.$$

Substituting this result in the first expression of (6.11) for the kinetic energy K, we arrive at the result

$$K = \sum_{i=1}^{N} \frac{m_i v_i'^2}{2} + \mathbf{V} \cdot \sum_{i=1}^{N} m_i \mathbf{v}'_i + \sum_{i=1}^{N} \frac{m_i V^2}{2}.$$

Note that in the system (C), the total momentum is zero,

$$\sum_{i=1}^{N} m_i \mathbf{v}'_i = 0,$$

where $M = m_1 + m_2 + \ldots + m_N$. Thus, we obtain the important equation

$$K = K_c + \frac{MV^2}{2}, \tag{6.12}$$

where $\mathbf{V}$ is the speed of its center of mass at the laboratory system and K_c is the kinetic energy of the system relative to the center of mass (in the frame (C)). Usually we call K_c internal kinetic energy. The relation (6.12) shows that the internal kinetic energy can be separated from the kinetic energy of the translational movement of the body as a whole. The latter has the same form as the kinetic energy of a single particle.

The relation (6.12) is extremely useful for the practical calculation of K in many complicated cases. We will see this in several examples in the next section and exercises. However, the importance of this relation goes far beyond its usefulness. Let us remember that the energy considered here is mechanical and it can be called macroscopic energy. At the same time, we know that macroscopic bodies consist of microscopic components like molecules and atoms. The Eq. (6.12) explains why it is not necessary to take into account the internal movement of these elements when we intend to study the movements of bodies as a whole. The same consideration applies to macroscopic systems which consist of several parts, also macroscopic. For example, when studying the Earth's motion around the Sun, in the zero-order approximation we do not need to take into account the rotation of the earth itself or the movement of the Moon around the Earth. We simply consider the Earth as a single point-like body.

Exercise 1. Find a more general equation than (6.12) for the transformation of an inertial frame to another with relative velocity $\mathbf{u}$. Show that (6.12) appears as a particular case.

6.4 Applications of Energy Conservation

We will consider a few examples of using the law of conservation of energy, some of them will be given as exercises.

Example 1. A mass is sliding from a height h along a complicated trajectory (as shown in Fig. 6.3), without friction. Assuming that the initial velocity is zero, calculate the speed at the ground level.

The solution by Newton's second law is complicated because we do not know the shape of the trajectory. But, fortunately, energy conservation law enables us to get a simple solution. The energy of the particle of mass m is given by

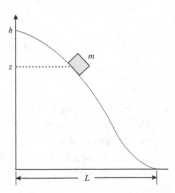

Fig. 6.3 The variation of potential energy is independent of the trajectory (Example 1)

$$E = \frac{mv^2}{2} + mgz = \text{constant},$$

where v is the particle speed in the altitude z. Thus, comparing the energies at the top and at the ground level, we obtain

$$\frac{mv^2}{2} = mgh$$

and this gives us the solution for the final velocity, $v = \sqrt{2gh}$, regardless of the shape of the trajectory.

Example 2. The same problem can be solved in the case of a wheel of mass m and radius R, assuming that the whole mass of the wheel is concentrated on its circumference. The friction is large and the sliding does not happen (Fig. 6.4).

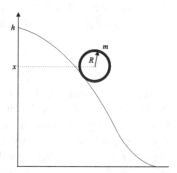

Fig. 6.4 Example 2. The same problem for a wheel (circular ring)

To solve this problem, we need a formula for the kinetic energy of the massive rolling circle, whose center moves with velocity v. Note that all points of the wheel have different velocities at the laboratory reference frame, (L).

Let us use the Eq. (6.12). Firstly we calculate the internal kinetic energy. Obviously, at the reference (C) the wheel is rotating around a static axis. As a result, all points of the wheel have the same speed in this reference frame. To find the value of the speed, we remember that the point where the wheel touches the ground (we can call it contact point) has zero velocity relative to the ground. As the center of the wheel has speed v relative to the ground, in the direction of the movement, the contact point should have speed v relative to the center of the wheel in the opposite direction to its movement. Again, all the points of the wheel have equal speed v in relation to the center of the wheel. Therefore,

$$K_c = \frac{mv^2}{2}.$$

At the same time, the kinetic energy associated with the movement of the center of mass is given by $mV^2/2 = mv^2/2$. Finally, we have

$$K = K_c + \frac{mV^2}{2} = \frac{mv^2}{2} + \frac{mv^2}{2} = mv^2.$$

Using the law of energy conservation, we arrive at the result for the speed of the wheel at the end of the path,

$$mv^2 = mgh \qquad \Longrightarrow \qquad v = \sqrt{gh}. \tag{6.13}$$

It is easy to see that the speed is reduced compared to the sliding block, because part of the initial potential energy is spent for rotation.

Example 3. Calculate the escape velocity for a planet of mass M and radius R.

Solution. The escape velocity is defined as the minimum speed required for an object to travel out of the reach of the gravitational attraction of a planet or star, by escaping to infinity. It is supposed that the only force acting on the test body is the gravitational one.

As we already know, the potential energy of a test particle of mass m in the gravitational field generated by a planet or a star of mass M has the form

$$U(r) = -\frac{GMm}{r}$$

and is defined such that $U(\infty) = 0$. The criterion for a test particle to escape from the gravitational field is that its kinetic energy at infinity is not negative, that is, $K(\infty) \geq 0$, simply because the relationship $K < 0$ admits no physical interpretation. Therefore, the escape velocity v_2 is defined by the equality $E = K + U = 0$, because for $E < 0$ the particle will not get out to infinity and $E > 0$ means that the initial speed would not be the minimal possible one. Then we get

$$E = \frac{mv_2^2}{2} - G\frac{mM}{R} = 0,$$

that means $v_2 = \sqrt{2GM/R}$. This result can be compared to the velocity v_1 of a particle in circular orbit near the Earth's surface, defined by Newton's second law as

$$\frac{mv_1^2}{R} = \frac{GmM}{R^2} \qquad \Longrightarrow \qquad v_1 = \sqrt{\frac{GM}{R}} = \frac{v_2}{\sqrt{2}}.$$

If we want to calculate values for the Earth, we can use $g = \frac{GM}{R^2}$, and so

$$v_1 = \sqrt{gR} \cong \left(10\,\mathrm{m/s^2} \cdot 6.4 \cdot 10^6\,\mathrm{m}\right)^{1/2}$$

$$= \left(64 \cdot 10^6\,\mathrm{m^2/s^2}\right)^{1/2} = 8{,}000.00\,\mathrm{m/s} = 8\,\mathrm{km/s}.$$

Correspondingly, $v_2 = \sqrt{2}v_1 \approx 11.2 \cdot 10^3\,\mathrm{m/s} = 11.2\,\mathrm{km/s}$.

Example 4. A body of mass m has velocity $\mathbf{v}$ and another identical body is at rest. Calculate the internal kinetic energy of the system and compare the results with kinetic energy in the reference (L) obtained in two different ways:

(i) Direct approach, using additivity of the kinetic energy.
(ii) Using the Eq. (6.12).

Solution. The velocity of the center of mass is $\mathbf{v}/2$ and velocities of the bodies in the system (c) are

$$\frac{\mathbf{v}}{2} \quad \text{and} \quad -\frac{\mathbf{v}}{2}.$$

Correspondingly, the kinetic energy of the two bodies in the frame of the center of mass is given by

$$K_c = 2 \cdot \frac{m}{2}\left(\frac{v}{2}\right)^2 = \frac{mv^2}{4}.$$

In addition to the energy associated with the movement of the center of mass, we have

$$\frac{MV^2}{2} = \frac{2m}{2}\left(\frac{v}{2}\right)^2 = \frac{mv^2}{4}, \qquad \text{where} \qquad M = 2m,$$

such that, according to (6.12), $K = mv^2/2$. In the approach (i), obviously, the result is the same.

Exercises

1. Two bodies of masses m and $2m$ were thrown into the air and collided at the moment when their velocities were orthogonal and had absolute values $2v$ and v, respectively. After an instant collision, they continued the movement together forming a single body.
(a) Calculate the magnitude of the velocity immediately after the collision.
(b) What was the change of velocity of the center of mass due to the collision?
(c) Calculate the change in kinetic energy in this collision. Perform calculations in both reference (L) and reference (C).

Answer:(c) $\quad \Delta K = K_{ini} - K_{fin} = -\frac{5}{3}mv^2.$

2. What is the work of the centripetal force on a particle of mass m in a circular movement of radius R and velocity v? What kind of force (dissipative, potential, solenoidal) the centripetal one can be?

Answers: $W = 0$. The force may be either solenoidal or potential.

3. A block of mass M is sliding on a surface as shown in Fig. 6.3, from the initial height h, and reaches the lower surface after covering the horizontal distance L. Show that the work of the friction force (whose coefficient is μ) is given by $-\mu MgL$, regardless of the form of the surface.

4. For the previous problem, calculate the speed of the block at the end of the process, assuming that it leaves from rest, and establish the criterion by which it reaches the level of height zero.

Solution. Consider Cartesian coordinates, such that the trajectory is given by $y(x)$, with $y = h$, $x = 0$ at the beginning and $y = 0$, $x = L$ at the end of the path. To make the descent of the mountain until the end, the kinetic energy should be positive in all instants of a movement, i.e., the criterion is given by $Mg(h - y) > \mu Mgx$. The answer for the velocity is $v^2 = 2g(h - \mu L)$.

5. Consider an inelastic collision between two blocks of masses m and M, where the first of the two blocks was initially at rest and another one had kinetic energy given by $\mathscr{E}$.

(a) For which relation between the masses m and M the loss of mechanical energy in the collision process reaches its minimum and for which relation it reaches its maximum?

(b) In the case where the masses of the blocks are equal $M = m$ and they have, initially, velocities $\mathbf{v}_1$ and $\mathbf{v}_2$, with identical modulus, $v_1 = v_2$, which are the angles between the two speeds for the loss of mechanical energy to be minimum and maximum? Try to solve this last part without performing any calculation.

Answer: *(a)* If $m/M = k$, the loss of energy is

$$\Delta \mathscr{E} = \mathscr{E} - \mathscr{E}_{fin} = \frac{k}{1+k} \mathscr{E},$$

so that when $k \to 0$, we have $\Delta \mathscr{E} \to 0$ and when $k \to \infty$, we have $\Delta \mathscr{E} \to \mathscr{E}$, so that $\mathscr{E}_{fin} \to 0$.

(b) The loss of energy is maximum when the angle between the initial velocities is given by π.

6. Derive the dynamical equations and calculate the accelerations of the blocks for the system shown in Fig. 6.5 by using the following approaches:
(a) Direct application of Newton's Laws and
(b) Conservation of energy. There is no friction and the pulleys have negligible mass.

Answer: $a_M = \dfrac{2m - 4M}{4M + m} g = -2a_m.$

7. A bar of uniform mass M and length L spins around an axis perpendicular to the bar, with an angular velocity ω. The axis of rotation passes through one of the ends of the bar. Calculate its kinetic energy in the reference of the laboratory and at the center of mass.

Answer: Using integration over the points of the bar, we obtain $K_{lab} = \frac{1}{6}ML^2\omega^2$ and $K_C = \frac{1}{24}ML^2\omega^2$. Note that the kinetic energy of a point mass M which runs a circle of radius $L/2$ with an angular velocity ω is exactly the difference of the values above, $\frac{1}{8}ML^2\omega^2$, confirming the formula (6.12).

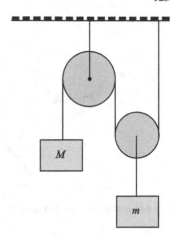

Fig. 6.5 Illustration for Exercise 6

8. A bar of uniform mass M and length L has a point particle of mass M attached to its end. The system rotates around the axis perpendicular to the bar, that passes through the center of mass of the system, with an angular velocity ω. Calculate the kinetic energy of the bar.

Answer : $\dfrac{5}{48} M L^2 \omega^2$.

9. Consider a circular ring of mass M and radius R, with all the mass concentrated on its circumference. The ring is spinning without slipping on the inclined plane which makes an angle θ with the horizontal direction. The initial speed of the wheel is zero. Calculate the velocity and acceleration of the wheel as a function of time.

Hint. Use the result of Example 2 in Sect. 6.4. Write $v = v(l)$, where $l = h/\sin\theta$ is the distance traveled by the ring, and take its time derivative to find the acceleration.

Answer : $\quad v = \dfrac{gt}{2} \sin\theta , \qquad a = \dfrac{g}{2} \sin\theta .$

10. Consider the previous problem of wheel, which spins along a horizontal plane without slipping. The plane moves in the horizontal direction, with an acceleration $a_0 = const$. Calculate the acceleration of the wheel.

Hint. Use the result of the previous problem and take into account the equivalence of forces of gravity and inertia. In a non-inertial reference frame linked to the accelerated plane, the wheel is a subject to the effective gravity given by $\mathbf{g}_{ef}$ making an angle θ with $\mathbf{g}$, where $g_{ef} = \sqrt{g^2 + a_0^2}$ and $\tan\theta = a_0/g$. Thus, the problem can be solved by analogy with the movement of the inclined plane (in physics, we use the term "analog model" to characterize this situation).

Answer: $\quad a = a_0/2.$

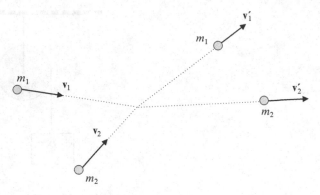

Fig. 6.6 Elastic collision between two particles

6.5 Elastic Collisions Between Two Particles

As an important example of the application of conservation laws of energy and momentum, consider the elastic collision between two particles. The term "elastic collision" means that both conservation laws apply at the same time. A good example of elastic collision is a collision between two billiard balls. These balls are made of a very hard material, with low friction between them. Therefore, the energy loss at the moment of the collision is very small and the process can be seen as an elastic collision. In some exercises in the previous section we already considered inelastic collisions, where mechanical energy is not conserved in contrast to the elastic case.

We should emphasize that the problem of scattering (collision is a particular case of scattering) is of great importance in many areas, for example, in atomic physics and especially in high energy physics. In modern Physics the experiments which are most complicated and more expensive (e.g., the ones in CERN and FERMILAB) involve scattering of elementary particles in accelerators. In these experiments, particles like electrons or protons are accelerated to velocities very close to the light speed, so the appropriate theory is essentially relativistic. Moreover, quantum effects become important. For example, as a result of the collision of particles there can be some new generated particles, so the total rest mass of the resulting particles may increase considerably. In other words, the processes of scattering on modern accelerators is far beyond the reach of Classical Mechanics. But it is important to remember that Physics is a unique science and its different sections can be related in some way. For example, in any process of elastic scattering, being it for particles or fields, classical, relativistic or quantum, the laws of conservation of momentum and energy remain valid. Therefore, the study of the scattering processes in the classical framework is a necessary step for a better understanding of more complicated cases.

Consider a collision between two particles of masses m_1 and m_2, with initial velocities $\mathbf{v}_1$ and $\mathbf{v}_2$, respectively (See Fig. 6.6).

Our goal is to calculate the velocities of the particles, $\mathbf{v}_1'$ and $\mathbf{v}_2'$, after the elastic collision. It should be noted from the very beginning that we do not know the details of the interaction between the two bodies. It is therefore natural to expect some degree of uncertainty in the final result. However, despite our ignorance about the details of the collision, nothing prevents us from applying the laws of conservation of momentum and energy to describe the process. In the laboratory reference frame, (L), the law of conservation of the linear momentum is given by

$$m_1 \mathbf{v}_1 + m_2 \mathbf{v}_2 = m_1 \mathbf{v}_1' + m_2 \mathbf{v}_2', \tag{6.14}$$

and the law of conservation of energy can be written as

$$\frac{m_1 v_1^2}{2} + \frac{m_2 v_2^2}{2} = \frac{m_1 v_1'^2}{2} + \frac{m_2 v_2'^2}{2}. \tag{6.15}$$

In principle, one can try to solve these two equations directly. At the same time, there is a simpler alternative. Let us change to the reference center of mass of the system (C) and solve the problem in this reference system. After that, it is always possible to perform an inverse transformation to the reference of the laboratory.

The velocity of the center of mass of the system is

$$\mathbf{V} = \frac{m_1 \mathbf{v}_1 + m_2 \mathbf{v}_2}{m_1 + m_2} \tag{6.16}$$

and the velocities of the particles relative to the system (C) are given by

$$\mathbf{u}_1 = \mathbf{v}_1 - \mathbf{V} \qquad \text{and} \qquad \mathbf{u}_2 = \mathbf{v}_2 - \mathbf{V}. \tag{6.17}$$

It is easy to verify that the total momentum of the system in the reference frame (C) is null, so $m_1 \mathbf{u}_1 + m_2 \mathbf{u}_2 = 0$. The next step is to find the velocities of the particles in relation to the reference (C) after scattering, $\mathbf{u}_1'$ and $\mathbf{u}_2'$.

According to the law of conservation, the total momentum in the inertial reference system (C) remains zero after the scattering, such that $m_1 \mathbf{u}_1' + m_2 \mathbf{u}_2' = 0$. As a consequence, the relations between the initial velocities $\mathbf{u}_1$ and $\mathbf{u}_2$ and the final ones $\mathbf{u}_1'$ and $\mathbf{u}_2'$, will be the same,

$$\mathbf{u}_2 = -\frac{m_1}{m_2} \mathbf{u}_1, \qquad \mathbf{u}_2' = -\frac{m_1}{m_2} \mathbf{u}_1'. \tag{6.18}$$

An important detail is that the formulas (6.18) show that the direction of initial velocity of the first particle, $\mathbf{u}_1$, is the opposite to the direction of the initial velocity of the second particle, $\mathbf{u}_2$, and similarly for the final velocities. At the same time, these relations do not establish a link, for example, between the directions before and after the scattering, e.g., between $\mathbf{u}_1$ and $\mathbf{u}_1'$. The origin of this feature is the absence of privileged directions in the reference (C), where the total momentum is null.

Taking into account energy conservation, we can write

$$m_1 u_1^2 + m_2 u_2^2 = m_1 u_1'^2 + m_2 u_2'^2. \tag{6.19}$$

Then, from the Eqs. (6.18) and (6.19), we obtain the relation

$$\left(m_1 + \frac{m_1^2}{m_2}\right) u_1^2 = \left(m_1 + \frac{m_1^2}{m_2}\right) u_1'^2 ,$$

i.e., for the absolute values of the velocities we have $u_1 = u_1'$, and, according to (6.18), $u_2 = u_2'$. So, in the system (C), the absolute values of the speeds do not change. At the same time, since the total momentum is zero, as mentioned above, there is no preferred direction and therefore one can not predict the direction of $\mathbf{u}_1'$. In fact, this direction depends on the details of the scattering. One can say that the interaction between the particles breaks the isotropy that exists initially in the reference (C). In this reference, the linear momentum of the system is null and there is no privileged direction for the scattering of the particles. However, the details of the interaction during the collision always indicate such a particular direction, which can be parameterized by the unit vector $\hat{\mathbf{n}}$. Let us suppose that

$$\mathbf{u}_1' = \hat{\mathbf{n}} u_1$$

in the reference (C). In this case, according to (6.18),

$$\mathbf{u}_2' = -\frac{m_1}{m_2} \hat{\mathbf{n}} u_1 = -\hat{\mathbf{n}} u_2 .$$

The last two equations contain all possible information on the outcome of the collision in the reference of the center of mass (C). All the uncertainty of the result (coming from the unknown details of the collision) is accumulated in the arbitrariness of the unit vector $\hat{\mathbf{n}}$. As we have just shown, the conservation laws are not sufficient to determine $\hat{\mathbf{n}}$.

Now we are in a position to consider the collision in the reference (L). For this end we have to perform a small calculation. Remember we can write the velocities in the reference (C) as follows:

$$\mathbf{u}_1 = \mathbf{v}_1 - \mathbf{V} = \frac{m_1\mathbf{v}_1 + m_2\mathbf{v}_1 - m_1\mathbf{v}_1 - m_2\mathbf{v}_2}{m_1 + m_2} = \frac{m_2}{m_1 + m_2}\left(\mathbf{v}_1 - \mathbf{v}_2\right); \quad (6.20)$$

also

$$\mathbf{u}_2 = \frac{m_1}{m_1 + m_2}\left(\mathbf{v}_2 - \mathbf{v}_1\right). \quad (6.21)$$

We can introduce a new notation for the difference between the velocities of the particles before the collision,

$$\mathbf{v}_1 - \mathbf{v}_2 = \mathbf{v}.$$

Using this notation we can write

$$\mathbf{u}_1 = \frac{m_2}{m_1 + m_2}\mathbf{v} \quad \text{and} \quad \mathbf{u}_2 = -\frac{m_1}{m_1 + m_2}\mathbf{v}.$$

Finally, by adding the velocity of the center of mass, $\mathbf{V}$, we arrive at the result

$$\mathbf{v}_1' = \mathbf{u}_1' + \mathbf{V} = \frac{m_1\mathbf{v}_1 + m_2\mathbf{v}_2}{m_1 + m_2} + \frac{m_2\,v\hat{\mathbf{n}}}{m_1 + m_2},$$

$$\mathbf{v}_2' = \mathbf{u}_2' + \mathbf{V} = \frac{m_1\mathbf{v}_1 + m_2\mathbf{v}_2}{m_1 + m_2} - \frac{m_1\,v\hat{\mathbf{n}}}{m_1 + m_2}, \qquad (6.22)$$

where $v = |\mathbf{v}| = |\mathbf{v}_1 - \mathbf{v}_2| = |\mathbf{v}_1' - \mathbf{v}_2'|$. These formulas represent a final output of our consideration in the reference (L).

For a better understanding of the last result, consider the following question: Can we have arbitrary directions for the velocities $\mathbf{v}_1'$ and $\mathbf{v}_2'$? In order to get the answer, let us rewrite (6.22) in terms of momenta of the particles,

$$\mathbf{p}_1' = m_1\mathbf{V} + \hat{\mathbf{n}}\mu v, \quad \mathbf{p}_2' = m_2\mathbf{V} - \hat{\mathbf{n}}\mu v, \qquad (6.23)$$

$$\text{where} \quad \mu = \frac{m_1 m_2}{m_1 + m_2}$$

is the reduced mass of the system of two particles. Remember that the notion of reduced mass has been introduced in the general problem of two bodies. In general the problem of collision can be seen as a particular case of the problem of two bodies of Chap. 4, therefore it is not a surprise to meet the reduced mass μ here.

The expressions for the momenta (6.23) admit a clear interpretation. We can note that the terms in this formula $m_1\mathbf{V}$ and $m_2\mathbf{V}$ represent proportional parts of the total momentum, $\mathbf{P} = (m_1 + m_2)\mathbf{V}$. After the collision, this momentum is distributed to two particles proportionally to their masses. We can say that the particles have "forgotten" the "old distribution" of the total momentum of the system as a result of the collision. The second terms in the formulas (6.23) represent the momenta of the particles in the reference (C). Obviously, the sum of these terms is zero.

It is useful to consider a diagram of the scattering for the momenta. As an example, we have constructed a diagram to illustrate $\mathbf{p}_1'$, where we introduced the special notations for the first terms of the expressions (6.23), as

$$\mathbf{p}_1^* = m_1\mathbf{V}, \qquad \mathbf{p}_2^* = m_2\mathbf{V}.$$

The diagram is shown in Fig. 6.7, where the radius of the circle is μv and the final momentum is equal to

$$\mathbf{p}_1' = \mathbf{p}_1^* + \hat{\mathbf{n}}\mu v.$$

The unit vector $\hat{\mathbf{n}}$ can have all possible directions, which, taken together, form the circumference. We can see that the possible angles θ_1, to the direction of the vector $\mathbf{p}'$ (and correspondingly for the velocity $\mathbf{v}_1'$) can be defined through this diagram.

Naturally, if $\mu v > m_1 V$, the initial point of the vector $m_1\mathbf{V}$ is inside the circle. In this case, there are no limits or restrictions for the angle θ_1.

In contrast to this, if $\mu v < m_1 V$, one can note that $|\theta_1| \leq \theta_{1\text{max}}$, where $\theta_{1\text{max}}$ corresponds to the situation in which $\hat{\mathbf{n}} \perp \mathbf{p}_1'$ (see Fig. 6.8). So, for the maximum angle of our interest we have

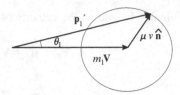

Fig. 6.7 Diagram for $\mathbf{p}_1'$

$$\sin \theta_{1\max} = \frac{\mu v}{m_1 V} = \frac{m_2 v}{(m_1 + m_2) V}. \tag{6.24}$$

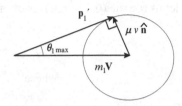

Fig. 6.8 Diagram of the limit
situation $\theta = \theta_{\max}$, when
$\mathbf{p}_1' \perp \hat{\mathbf{n}}$

Of course, the same diagram can be constructed for the particle of mass m_2, the modification being $\hat{n} \to -\hat{n}$. In this case, the restriction for θ_2 has the form $|\theta_2| < \theta_{2\max}$, where

$$\sin \theta_{2\max} = \frac{\mu v}{m_2 V} = \frac{m_1 v}{(m_1 + m_2) V}. \tag{6.25}$$

In Analytical Mechanics we can meet more sophisticated diagrams, describing more complicated scattering processes.

Exercises

1. Show that if $\mathbf{v}_1 \perp \mathbf{v}_2$ before an elastic collision between two particles of equal mass, the same is true after the collision, namely $\mathbf{v}_1' \perp \mathbf{v}_2'$.

2. Consider, without using the results of this section, a frontal collision, in which the directions of the velocity of the particles does not undergo any change, i.e., $\mathbf{v}_1 \| \mathbf{v}_1'$ and $\mathbf{v}_2 \| \mathbf{v}_2'$. Compare the results for the final velocities with the general formulas, considering the special case $\hat{\mathbf{n}} \| \mathbf{V}$.

3. Consider the particular case of elastic collision in which particle masses are equal, $m_1 = m_2 = m$, and one of the particles is initially at rest, $\mathbf{v}_1 = \mathbf{v}$ and $\mathbf{v}_2 = 0$. Show that the angle between the directions of the particles velocities after the collision is equal to $\pi/2$. Solve this problem in three different ways:

(*i*) Using the formula (6.23);

(ii) Using the formula (6.22);

(iii) Without using these formulas or the transition to the reference (C). The calculations must be performed directly through the relations (6.14) and (6.15).

Hint. The solution in the latter case is the simplest one.

4. Consider the diagram in Fig. 6.7. If the angle for the particle of mass m_2 is θ_{2max}, what is the angle of scattering for the particle m_1? May this angle be equal to θ_{1max}?

Solution. We will use the formula (6.23) as a starting point. The angle between the vectors $m_2 \mathbf{V}$ and $-\mu \hat{n} V$ is given by θ_{2max}. You can check that, by using Eq. (6.23) and geometrical considerations, the angle between the vectors $m_1 \mathbf{V}$ and $\mu \hat{n} V$ is equal to $\alpha = \frac{\pi}{2} + \theta_{2max}$. Using the fact that the vectors $m_1 \mathbf{V}$, $\mu \hat{n} V$ and $\mathbf{P}'_1$ form a triangle, we obtain

$$P'_1 = \left[m_1^2 V^2 + \mu^2 v^2 + 2m_1 \mu v V \cos \theta_{2max} \right]^{1/2}, \qquad \sin \theta_1 = \frac{\mu v}{P'_1} \cos \theta_{2max} .$$

Taking into account $\alpha > \pi/2$, the angle α can not be equal to the angle θ_{2max}.

5. Consider the frontal elastic collision between two bodies of masses m and M, where the body of mass M is initially at rest. Assuming that the initial kinetic energy of the body of mass m is given by E_0 and the final kinetic energy of the body of mass M is E (transferred energy), for which relation between the two masses the ratio E/E_0 reaches its maximum and minimum values? Compare your answers with the case of inelastic collision.

Answer : $\dfrac{E}{E_0} = \dfrac{4Mm}{(m+M)^2}$;

is maximal for $m = M$ and minimal for $M \ll m$ or $M \gg m$.

Chapter 7
Movement in a Potential Field: Oscillations

Abstract This chapter will be devoted to the important phenomenon of oscillations, which has numerous applications and generalizations in several areas of Physics. The small oscillations without friction manifest a property of universality, such that they can be treated as harmonic ones. More complicated cases include harmonic oscillations in the presence of damping force and external source. We shall present a few simple examples illustrating the treatment of multidimensional oscillators by means of normal modes.

7.1 Introductory Remarks

In this chapter we consider a particular important case of the dynamics of a single particle, with the emphasis on the motion which is confined to a certain region of space. In the first sections we consider the motion of a particle which is restricted to a single dimension. In this case the potential energy (or just the "potential") is a function of a single coordinate, $U(x)$, in the absence of dissipative forces, such that the total mechanical energy of the particle is constant. The law of energy conservation enables one to perform a general classification of the allowed types of motion, including the movements restricted or not to a finite region. A useful approach is to consider motions involving only small deviations from an equilibrium position. In this case, the solutions have a very special universal form. This approach is called harmonic oscillator. Next, we consider more complicated cases, in the presence of dissipative and external forces. In the last section, we discuss a few examples related to two-dimensional oscillations.

I.L. Shapiro and G. de Berredo-Peixoto, *Lecture Notes on Newtonian Mechanics*,
Undergraduate Lecture Notes in Physics, DOI 10.1007/978-1-4614-7825-6_7,
© Springer Science+Business Media, LLC 2013

7.2 Unidimensional Movement in a Fixed Potential

There are two possibilities for studying a one-dimensional motion in a fixed potential, namely one can use Newton's Second Law directly or rely on the energy conservation law. Mathematically the two approaches are equivalent.

Assuming that the potential energy of a particle of mass m is given by $U(x)$ and that its kinetic energy is given by the expression $K = mv^2/2$, one can apply conservation of energy and write

$$\frac{mv^2}{2} + U(x) = E = \text{const},$$

to study the motion of the particle. In general, for a fixed potential $U(x)$, the particle can perform very different movements, depending on the initial conditions, especially on the value of total energy E. We should note that the value of E is determined by the initial values of x and v. For example, if we define $x(t_0) = x_0$ and $v(t_0) = v_0$, the mechanical energy is

$$E = \frac{mv_0^2}{2} + U(x_0).$$

From another side, the values of x_0 and v_0 constitute the initial data for Newton's second law,

$$\ddot{x} = \frac{F(x)}{m} = -\frac{1}{m}\frac{dU}{dx}.$$

For that reason, when the energy E is given, it is necessary to fix only one of these quantities in order to have a set of complete initial data, to both x_0 and v_0 being included into E. Later in this section, we will treat this aspect with more details.

The use of relation $E = K + U = constant$ opens the way to classify the possible movements for a certain potential $U(x)$. The idea behind this classification is quite simple, and it can be summarized in the fact that the kinetic energy

$$K = E - U(x) = \frac{mv^2}{2}$$

can not be negative. For example, if we just put on the same plot the curves $U(x)$ and the straight horizontal line $E = constant$, it is easy to see that only those regions where $E - U(x) \geq 0$ are permitted for the motion of the particle (or the mechanical system, in general). For any other type of points, where $E - U(x) < 0$, the presence of the particle is forbidden. This classification naturally depends on the value of E. Typically, the larger the energy E, the larger the area where the particle may be present.

In the diagram shown in Fig. 7.1, we can recognize the following features of motion, according to the value of the energy E:

1) For $E = E_0$, the kinetic energy is identically zero. It is only allowed for the particle to remain static at the point $x = x_0$, where the potential has its minimum.

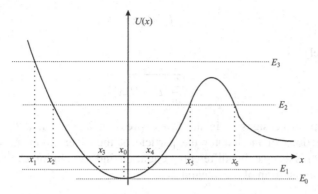

Fig. 7.1 The possible positions of a particle moving in the potential $U(x)$ depend on the value of its energy E

2) For $E = E_1$, the movement is confined in the region $x_3 \leq x \leq x_4$. In this case, the movement is called limited (or finite). The system can be also classified as *bounded*. The last means that the points with values of x which are arbitrarily large $(x \to \infty)$ are not available for the particle.

The extreme points of the intervals allowed for the movement are called *turning points*. For example, if the energy is $E = E_1$, we have two turning points, namely x_3 and x_4. When the particle approaches the point of return, its potential energy approaches E, therefore its kinetic energy, $K = E - U(x)$, becomes small until it reaches zero when the particle reaches such a point. At this point the particle stops (because $K = 0$), and its acceleration,

$$\ddot{x} = a = -\frac{F}{m} = -\frac{1}{m}\frac{dU}{dx},$$

since it is different from zero, immediately takes the particle from instantaneous rest to move in the opposite direction. For this reason, it is called the turning point.

3) For $E = E_2$, the values $x \geq x_6$ and $x_2 \leq x \leq x_5$ are allowed with a total of three return points. In principle, for this level of energy the motion of the particle can be unlimited, but this will depend on the starting position. If at the time instant t_0 the particle is in the region $x \geq x_6$, the movement is unlimited. On the other hand, if at the instant t_0 the particle is inside the region $x_2 \leq x \leq x_5$, the motion will be limited because the particle can not cross the potential barrier between the points x_5 and x_6.[1]

4) For $E = E_3$, there is only one return point and the movement is infinite.

As was already mentioned, the law of conservation of energy can be used directly to find the motion of the particle. The law of energy conservation can be written as

---

[1] It is worth noting that this restriction is only valid in classical mechanics. In quantum physics, the passage through a potential barrier is possible, in a phenomenon called "quantum tunneling".

$$\frac{m}{2}\dot{x}^2 + U(x) = E$$

or, alternatively, as

$$\dot{x} = \pm\sqrt{\frac{2}{m}\left[E - U(x)\right]}. \tag{7.1}$$

The two possible signs $\pm$ in the last expression indicate that there are two possible directions of motion for the particle at the given point x. Mathematically, Eq. (7.1) is a separable differential equation and its solution can be obtained immediately in the form of the integral

$$\pm\sqrt{\frac{m}{2}}\int_{x_{\text{ini}}}^{x}\frac{dx}{\sqrt{E - U(x)}} = t - t_0. \tag{7.2}$$

The last formula allows, among other things, to calculate the time elapsed in the movement between the two turning points, say, x_2 and x_5, for the example illustrated in Fig. 7.1, where $E = E_2$. Let us define the moments in which the particle reaches the return points as t_2 and t_5, so that $x(t_2) = x_2$ and $x(t_5) = x_5$. Using the Eq. (7.2), we get the following expression for the interval of time:

$$t_5 - t_2 = \sqrt{\frac{m}{2}}\int_{x_2}^{x_5}\frac{dx}{\sqrt{E - U(x)}}.$$

The motion backward takes exactly the same time, until the particle returns to the point x_2, and the cycle repeats *ad infinitum*.[2] We can note that the limited movement of the particle in a fixed potential has oscillatory form, whose period is given by

$$T = 2(t_5 - t_2) = \sqrt{2m}\int_{x_2}^{x_5}\frac{dx}{\sqrt{E - U(x)}}. \tag{7.3}$$

Finally, it is interesting to note one more relationship between the conservation of energy,

$$E = \frac{m\dot{x}^2}{2} + U(x) = \text{const.}, \tag{7.4}$$

and Newton's Second Law. Taking the time derivative of (7.4), we find the equation

$$m\dot{x}\ddot{x} + \frac{dU}{dx}\dot{x} = 0.$$

---

[2] The real meaning of this is that the repetition of cycles will not end in a finite time. In practice, there will always be dissipative forces, and there will be a moment at which the particle stops at some point near the equilibrium position x_0.

Dividing by $\dot{x}$, one can get Newton's Second Law. Mathematically, this means that (7.4) represents the first integral of motion for the differential equation of second order

$$m\ddot{x} = -U', \qquad \text{where} \qquad U' = \frac{dU}{dx}. \tag{7.5}$$

Indeed, one can also start from the second Law (7.5) and arrive at the conservation law (7.4) by means of the procedure called "lowering the order of differential equation". We have discussed this procedure for a particular example in Chap. 3 and leave it as an exercise for the reader to check that the same also works in the general case. From this perspective it becomes clear what is the advantage of using the energy conservation. In this case one simply has one less integral to take.

Exercise 1. Solve the same one-dimensional motion problem and obtain the Eq. (7.3) directly using Newton's Second Law. This approach is more complicated because it involves an extra integration.

7.3 Harmonic Oscillator

The practical use of the formula (7.2) and, in general, derivation of the motion for the particle in the potential $U(x)$, depends on our ability to evaluate the integral (7.2). In many cases, depending on the form of the potential $U(x)$, the solution of this problem can be very difficult. However, there is a special case in which the details of the potential are not relevant and the solution can be determined in some universal general way. Such a simplification occurs in a case when a particle performs small oscillations about the minimal point of the potential, as it is shown in Fig. 7.2.

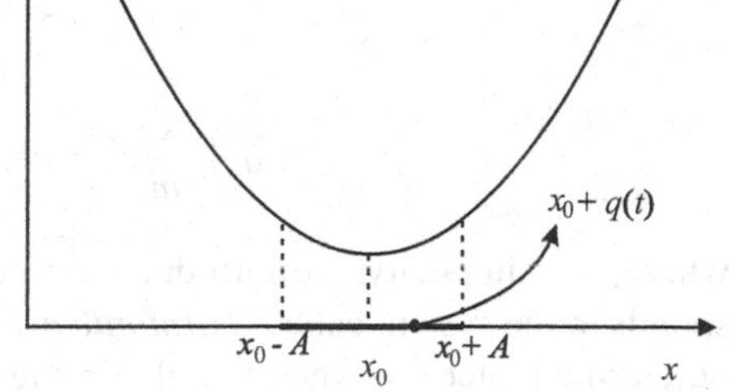

Fig. 7.2 Small oscillations around the equilibrium point. The amplitude, A, is by definition a small amount, $A \ll l$ (l is some typical length characterizing the dimensions of the oscillating system)

At the point of the minimum of the potential, x_0, we have

$$U'(x_0) = \left.\frac{dU}{dx}\right|_{x=x_0} = 0.$$

It is clear that for the case $x(t_0) = x_0$ and $v(t_0) = 0$, the particle will remain at the point x_0 forever. We consider a small deviation, $q(t)$, of the point x_0. The position of the particle under these conditions, at the time instant t is given by the expression

$$x = x_0 + q(t),$$

where $|q(t)|$ is considered to be very small. This means that we can consider only terms of first non-trivial order in $q(t)$ and discard the terms of higher orders in $q(t)$ and its time derivatives.

The next point is to rewrite Newton's Second Law in terms of $q(t)$. For this purpose, according to our previous discussion, one can consider the law of energy conservation and replace $x = x_0 + q(t)$. In the kinetic energy term we have $\dot{x} = \dot{q}$ and therefore $K = m\dot{q}^2/2$. The potential term must be expanded around x_0, according to

$$U(x) = U(x_0) + \left.\frac{dU}{dx}\right|_{x_0} q + \frac{1}{2}\left.\frac{d^2U}{dx^2}\right|_{x_0} q^2 + \dots . \tag{7.6}$$

Taking into account the identity $U'(x_0) = 0$, we can conclude that the first significant term is quadratic in $q(t)$. Hence, let us consider

$$U(x_0 + q) = \text{const} + \frac{1}{2}kq^2 + \mathcal{O}(q^3), \quad \text{where} \quad k = \left.\frac{d^2U}{dx^2}\right|_{x=x_0}.$$

The terms of the third and higher orders, $\mathcal{O}(q^3)$, can be neglected, because $q(t)$ is assumed to be very small. Then in the quadratic approximation (bilinear) we have

$$K + U = \frac{m\dot{q}^2}{2} + \frac{kq^2}{2} = E = \text{const}.$$

Taking the time derivative, we obtain

$$(m\ddot{q} + kq)\,\dot{q} = 0,$$

i.e.,

$$\ddot{q} + \frac{k}{m}q = \ddot{q} + \omega_0^2 q = 0, \tag{7.7}$$

where, for simplicity, we introduce the notation $\omega_0^2 = k/m$. The Eq. (7.7) corresponds to the system called *harmonic oscillator*. The origin of this expression will be clarified later on, when we obtain the solution of (7.7). As we have seen, the harmonic oscillator is a universal description of any system (with few exceptions) subject to small fluctuations in the neighborhood of a point of a stable equilibrium.

A relevant observation is that the Eq. (7.7) could be obtained in a more simple and intuitive way. Obviously, (7.7) means that the force $F = m\ddot{q}$ is proportional to the deviation of the particle (or a generic system) from the equilibrium position. In most cases, for small deviations the force will be proportional to the actual deviation according to (7.7). In practice, there may be dissipative contributions, e.g., frictional forces or air resistance. In this case, the Eq. (7.7) should be modified. We will study this and some other generalizations of (7.7) in the next sections, and we will now focus our attention on the solution of the equation (7.7).

Mathematically, (7.7) is a homogeneous ordinary differential equation of second order. To solve this and many other similar equations, it is necessary to account for some results of the theory of ordinary differential equations. The relevant mathematical information can be formulated in the form of the following theorem:

Theorem 1. *For a homogeneous linear equation,*

$$\ddot{q} + 2\gamma\dot{q} + \omega_0^2 q = 0, \tag{7.8}$$

there is always a fundamental system of solutions. This system consists of two particular solutions, $q_1(t)$ and $q_2(t)$, satisfying the following conditions:

- *Linear independence, whose criterion is the following[3]:*

$$W(t) = \begin{vmatrix} q_1(t) & q_2(t) \\ \dot{q}_1(t) & \dot{q}_2(t) \end{vmatrix} \neq 0. \tag{7.9}$$

The function $W(t)$ is called determinant of Wronski, or Wronskian.
- *The general solution of (7.8) can be expressed as a linear combination,*

$$q(t) = C_1 q_1(t) + C_2 q_2(t),$$

where C_1 and C_2 are arbitrary integration constants.

Note that this theorem is valid not only for an equation with constant coefficients, but also for the ones with variable coefficients, where $\gamma = \gamma(t)$ and $\omega_0^2 = \omega_0^2(t)$. Generally, the solution of mechanical problems may require the general case. But in this book, we confine ourselves to the simplest case with constant coefficients. For now, it will be convenient to consider $\gamma = 0$ and then we will study the case where $\gamma \neq 0$.

Taking into account the above theorem, all we should do, in practice, is to find two independent solutions, $q_1(t)$ and $q_2(t)$. A natural choice is the exponential, $q(t) = e^{\lambda t}$. For this function, we have $\ddot{q} = \lambda^2 q$. Substituting this into the Eq. (7.7), we find

---

[3] Two functions $q_1(t)$ and $q_2(t)$ are linearly independent if $B_1 q_1(t) + B_2 q_2(t) \equiv 0$ implies $B_1 = B_2 = 0$. We leave it as an exercise to show that if this criterion is not satisfied, the Wronskian $W(t)$ is identically zero. The demonstration of the sufficiency of $W(t) \neq 0$ for linear independence requires greater efforts, but can be found in many courses of differential equations (see, e.g., [4,7]).

$$\lambda^2 + \omega_0^2 = 0 \quad \text{i.e.,} \quad \lambda = \pm i\omega_0. \tag{7.10}$$

where i is the imaginary unit number, $i^2 = -1$. The last formula implies the existence of two complex solutions, $e^{i\omega_0 t}$ and $e^{-i\omega_0 t}$. At the same time, our intention is to use the Theorem about fundamental system of solutions for the Eq. (7.7) and hence we need to find two linearly independent real solutions. We know that

$$e^{\pm i\omega_0 t} = \cos \omega_0 t \pm i \sin \omega_0 t.$$

Taking into account the linearity of the Eq. (7.7), it is obvious that the real and imaginary parts of this solution, separately, are solutions of this equation. Thus, we have found two linearly independent real solutions, namely

$$q_1(t) = \cos \omega_0 t \quad \text{and} \quad q_2(t) = \sin \omega_0 t. \tag{7.11}$$

In accordance to the Theorem about fundamental system, a general solution of equation (7.7) has the form

$$q(t) = C_1 \cos \omega_0 t + C_2 \sin \omega_0 t, \tag{7.12}$$

where the coefficients C_1 and C_2 are directly related to the initial data. The functions cos and sin are examples of harmonic functions. For this reason the oscillations described by the Eq. (7.11) are called harmonic oscillations. Remember that the Eq. (7.7) itself is called the equation of the harmonic oscillator.

It proves convenient to present the general solution as

$$\begin{aligned} q(t) &= C_1 \cos \omega_0 t + C_2 \sin \omega_0 t \\ &= \sqrt{C_1^2 + C_2^2} \left(\frac{C_1}{\sqrt{C_1^2 + C_2^2}} \cos \omega_0 t + \frac{C_2}{\sqrt{C_1^2 + C_2^2}} \sin \omega_0 t \right) \\ &= \sqrt{C_1^2 + C_2^2} \cos(\omega_0 t + \varphi), \end{aligned} \tag{7.13}$$

where the angle φ is chosen so that

$$\cos \varphi = \frac{C_1}{\sqrt{C_1^2 + C_2^2}}, \quad \sin \varphi = - \frac{C_2}{\sqrt{C_1^2 + C_2^2}}.$$

The quantity $A = \sqrt{C_1^2 + C_2^2}$ is called *amplitude* of the oscillation. The solution can be rewritten in the standard form as

$$q(t) = A \cos(\omega_0 t + \varphi), \tag{7.14}$$

where ω_0 is the oscillation *frequency* and φ is called the initial *phase* of the oscillation. By definition, the amplitude is positive, $A > 0$. Typically, the phase is chosen so that $0 \le \varphi \le 2\pi$, or $-\pi \le \varphi \le \pi$, depending on the convenience. For the harmonic oscillator, the frequency of oscillations is independent of the amplitude.

As the reader will see after trying to solve Exercise 4, it is very difficult to find another example of the potential possessing this property.

One has to note that there is a big difference between the parameters ω_0 from one side and A, φ from the other. The magnitude of the frequency, ω_0, depends on the dynamical equation. In other words, it is a fundamental property of the given oscillator. This parameter does not depend on the particular movement or, equivalently, on the choice of initial data. In contrast, to determine the values A and φ (or C_1 and C_2), it is necessary to use the initial data.

Let us define these constants for $q(0) = q_0$ and $\dot{q}(0) = v_0$. For this end, one can write

$$\dot{q}(t) = \frac{dq}{dt} = -A\omega_0 \sin(\omega_0 t + \varphi).$$

Using the initial data specified above, we arrive at

$$A \cos \varphi = q_0, \qquad -A\omega_0 \sin \varphi = v_0. \tag{7.15}$$

Consequently, $q_0^2 \omega_0^2 + v_0^2 = A^2 \omega_0^2$, then

$$A = \sqrt{q_0^2 + \frac{v_0^2}{\omega_0^2}}.$$

Finally, the relations

$$\sin \varphi = -\frac{v_0}{A\omega_0}, \qquad \cos \varphi = \frac{q_0}{A}$$

enable one to define φ.

Let us calculate the mechanical energy of the oscillator and check explicitly that it is constant. Using the results given above, we obtain

$$K = \frac{mv^2}{2} = \frac{m\dot{q}^2}{2} = \frac{mA^2\omega_0^2}{2} \sin^2(\omega_0 t + \varphi),$$
$$U = \frac{kq^2}{2} = \frac{kA^2}{2} \cos^2(\omega_0 t + \varphi). \tag{7.16}$$

Remember that $m\omega_0^2 = k$ and we can conclude our checking,

$$E = K + U = mA^2\omega_0^2 = \text{const.} \tag{7.17}$$

As expected, the energy of the harmonic oscillator is constant. It is proportional to the square of the amplitude and frequency, and is independent of the phase φ.

Example 1. For a body of mass m hanging on the end of a spring with elastic constant k, we have, in the equilibrium position, $mg = -kx_0$, so that $x_0 = -mg/k$,

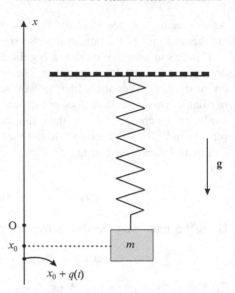

Fig. 7.3 Small oscillations around the point of equilibrium (Example 1)

where x_0 is the deformation that the spring undergoes (See Fig. 7.3). Calculate the frequency of small oscillations.

Solution. For small deviations from the equilibrium position, $x = x_0 + q$, we will use Hooke's law, $F = -kq$. Then Newton's second law, $m\ddot{q} = -kq$, is exactly the Eq. (7.7), and so we can use the procedure discussed above to find the solutions. The frequency of oscillations (number of oscillations per unit of time) is given by (both f and ω_0 are called frequency)

$$ f = \frac{\omega_0}{2\pi} = \frac{1}{2\pi}\sqrt{\frac{k}{m}}. $$

Observation. In this case, it is not important that the oscillations are small, because the potential energy depends quadratically of the deviation q. To understand the importance of this condition, it should be noted that for any real spring, the law of Hooke is only valid for sufficiently small deformations. For larger values, the deformation of the spring becomes irreversible and Hooke's law can not be applied.

Example 2. Consider a mathematical pendulum, i.e., the system shown in Fig. 7.4. The word "mathematical" means the following idealizations:
(i) The wire has mass equal to zero, such that the entire mass of the pendulum is concentrated at the end point;
(ii) The wire has absolute rigidity, i.e., its length is constant and does not depend on the strength.

The task is to consider small oscillations in a fixed plane and calculate the frequency and period of oscillations.

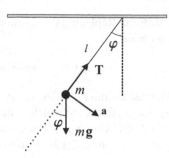

Fig. 7.4 Small oscillations
of mathematical pendulum
around the point of equilibrium (Example 2)

Solution. Applying Newton's Second Law,

$$\mathbf{T} + m\mathbf{g} = m\mathbf{a},$$

and using $\mathbf{a} \perp \mathbf{T}$ (because the wire does not deform), we obtain

$$ma = -mg \sin \varphi,$$

where the angle φ is measured positive in the counterclockwise direction. As far as
the trajectory of the particle is circular, we can write the known relationship between
the linear acceleration and the angular one, $a = l\ddot{\varphi}$ and obtain

$$\ddot{\varphi} = -\frac{g}{l} \sin \varphi. \tag{7.18}$$

For $|\varphi| \ll 1$ (in radians), one can expand $\sin \varphi$ into Taylor series around $\varphi = 0$,
according to

$$\sin \varphi = \varphi - \frac{1}{3!} \varphi^3 + \frac{1}{5!} \varphi^5 + \ldots \approx \varphi,$$

where the last approximation is valid only for very small angles. Thus, we can write
the equation of the harmonic oscillator,

$$\ddot{\varphi} = -\omega_0^2 \varphi,$$

where $\omega_0 = \sqrt{g/l}$. To determine the solution one just needs to use the formulas considered previously in this section. The period of oscillations is proportional to $1/\omega_0$,

$$T = \frac{1}{f} = \frac{2\pi}{\omega_0} = 2\pi \sqrt{\frac{l}{g}}.$$

This formula can be used to determine the value of g with reasonable accuracy by
experimental observation of oscillations.

There are many other examples of mechanical systems which can be described
as a harmonic oscillator in the regime of small oscillations.

Example 3. A harmonic oscillator with frequency ω_0 has an initial velocity given
by $\dot{q}(0) = v_0 < 0$ and its initial position corresponds to the equilibrium $q(0) = 0$.
Find the amplitude and the phase of the oscillations.

Solution. Replacing the solution (7.14) in the conditions $\dot{q}(0) = v_0$ and $q(0) = 0$, we obtain

$$\dot{q}(0) = -A\omega_0 \sin\varphi = v_0, \qquad q(0) = A\cos\varphi = 0. \tag{7.19}$$

The last formula implies $\cos\varphi = 0$. This means that the phase is equal to $\pi/2$ or to $-\pi/2$. To decide which of these values is correct, just note that, according to the initial condition $v_0 < 0$, we should have $\sin\varphi > 0$, so $\varphi = \pi/2$. Using the first initial condition, (7.19), we obtain for the amplitude

$$A = \frac{|v_0|}{\omega_0}.$$

Exercises

1. Consider again Example 2. Make the expansion of the potential energy $U(\varphi)$ around $\varphi = 0$ and derive the same Eq. (7.18) which was already obtained by using Newton's Second Law.

2. Using the results of Example 2, without expanding the series in φ, find the points of return for the finite motion and calculate the period of non-harmonic oscillations.

Solution. Writing the expression for the mechanical energy,

$$E = K + U(\varphi) = \frac{m\dot{\varphi}^2 l^2}{2} + 2mgl \sin^2\left(\frac{\varphi}{2}\right),$$

it is possible to establish the return points, since they correspond to the zero value of kinetic energy, $\pm\varphi_0$, where $\varphi_0 = 2\arcsin\sqrt{\frac{E}{2mgl}}$. The result for the period can be taken from the general formula (7.3),

$$T = 2l\sqrt{2m} \int_0^{\varphi_0} \frac{d\varphi}{\sqrt{E - 2mgl \sin^2\frac{\varphi}{2}}} = \sqrt{\frac{32l^2 m}{E}} F\left(\frac{\varphi_0}{2}, \sqrt{\frac{2mgl}{E}}\right),$$

where $F(\varphi, k)$ is the elliptic integral of the first type [12] (p. 860).

3. Consider the case of the potential shown in Fig. 7.1. For the specific value of energy $E = E_1$, what would be the speed of the particle at the points x_3 and x_4? What are the names of these points? Which are the signs of accelerations experienced by the particle at these points? Which quantity changes its sign at these points? Answer (if possible) using only (a) Newton's second law; (b) Energy conservation law.

4. Using the formula (7.3), verify that for the potential $U(x) = kx^2/2$ the period of oscillations does not depend on the energy. The first step for this demonstration is to find the turning points for a given value of the energy. As a much more complicated task, try to find another example of a potential which has the same feature.

Observation. The solution of the latter task is quite difficult. By now, only two different solutions for the potential are known, and one of them is the harmonic oscillator. A theoretical background of this problem can be found in the book [21].

5. Using (7.3), calculate the period of oscillations for the potential

$$U(x) = \frac{kx^2}{2} + \frac{\gamma x^4}{4},$$

where the coefficient γ is very small. To begin with, use considerations based on the dimensionality to define what it means for γ to be small. Explain the results in terms of the magnitude of the coordinates of turning points. In the formula (7.3), make the expansion in power series of γ to the first order and calculate the integral.

Solution (partial). A small value of γ means that the magnitude of the quartic term in the expression for the potential energy is much smaller than the quadratic term and also much smaller than the total mechanical energy. If the point of maximum deviation from equilibrium is $x = A$, small γ means $\gamma A^4 \ll kA^2$ and also $\gamma A^4 \ll E$.

We proceed now to the calculation of the period, using Eq. (7.3),

$$T = \sqrt{2m} \int\limits_{x_2}^{x_1} \frac{dx}{\sqrt{E - U(x)}},$$

where x_1 and x_2 are the roots of the biquadratic equation $kx^2/2 + \gamma x^4/4 = E$. Let us expand these roots to first order in γ. The biquadratic equation has the solutions

$$x_{1,2}^2 = \frac{1}{\gamma}\left(\sqrt{k^2 + 4\gamma E} - k\right). \tag{7.20}$$

Note that the other pair of roots is complex and hence unphysical. Thus, we obtain

$$x_{1,2} = \pm\sqrt{\frac{2E}{k}} \mp \left(\frac{E}{k}\right)^{3/2} \frac{\gamma}{k\sqrt{2}} + O(\gamma^2). \tag{7.21}$$

To calculate the period, we perform the expansion

$$T = \sqrt{2m} \int\limits_{x_1}^{x_2} \frac{dx}{\sqrt{E - kx^2/2 - \gamma x^4/4}} \tag{7.22}$$

$$= \sqrt{2m} \int\limits_{x_1}^{x_2} \frac{dx}{\sqrt{E - kx^2/2}} + \frac{\sqrt{2m}}{8} \int\limits_{x_1}^{x_2} \frac{x^4 \gamma dx}{(E - kx^2/2)^{3/2}} + \cdots.$$

For the last integral in the expansion above, it is convenient to make the change of variable $u = \sqrt{\frac{k}{2E}}x$. As a consequence, we arrive at the integral

$$I = \int_{u_1}^{u_2} \frac{u^4 du}{(1-u^2)^{3/2}} = \int_{\theta_1}^{\theta_2} \frac{\sin^4 \theta}{\cos^2 \theta} d\theta = \int_{\theta_1}^{\theta_2} \frac{d\theta}{\cos^2 \theta} - 2\int_{\theta_1}^{\theta_2} d\theta + \int_{\theta_1}^{\theta_2} \cos^2 \theta d\theta.$$

It is easy to show that $\theta_{1,2} = \pm \frac{\pi}{2} \mp \frac{\sqrt{E\gamma}}{k}$. By using these values, we obtain

$$T = 2\pi\sqrt{\frac{m}{k}} + \frac{\sqrt{2m}}{4}\gamma\tan\left(\frac{\pi}{2} - \frac{\sqrt{E\gamma}}{k}\right) + O(\gamma^{3/2}).$$

Using the identity $\tan(\pi/2 - \alpha) \equiv \cot\alpha$, and expanding up to the first order in γ, we arrive at the final result

$$T = 2\pi\sqrt{\frac{m}{k}} + \sqrt{\frac{mk^2\gamma}{8E}} + O(\gamma^{3/2}).$$

6. Verify that the criterion of linear independence, $W(t) \neq 0$, is satisfied for the functions (7.11).

7. A harmonic oscillator with frequency ω_0 has initial velocity given by $\dot{q}(0) = v_0 > 0$ and initial position $q(0) = q_0 > 0$. Find the amplitude and the phase of oscillations.

Answer:

$$q = A\cos(\omega_0 t + \varphi), \qquad A = \sqrt{q_0^2 + \frac{v_0^2}{\omega_0^2}}, \qquad \varphi = -\arctan\left(\frac{v_0}{\omega_0 q_0}\right).$$

8. A harmonic oscillator with angular frequency ω_0 and mass m has total mechanical energy E and initial velocity given by $\dot{q}(0) = v_0 < 0$. Find the amplitude of the oscillations and the possible values of the phase φ.

Answer:

$$A = \sqrt{\frac{2E}{m\omega_0^2}}, \qquad \varphi_1 = -\arcsin\left(v_0\sqrt{\frac{m}{2E}}\right), \qquad \varphi_2 = \pi - \varphi_1.$$

9. Consider the solution (7.12) for a harmonic oscillator with mass m.

(a) Define the integration constants, $C_{1,2}$, in terms of initial data, q_0 and v_0.
(b) Express the constants C_1 and C_2 in terms of the initial momentum p_0 and the mechanical energy E. Why does the problem have two solutions?
(c) Consider the two independent solutions, $q_1(t)$ and $q_2(t)$ from part (b). What is the relationship between them?

Answers:

(a) $C_1 = q_0$, $C_2 = \dfrac{v_0}{\omega_0}$;

(b) $C_1 = \pm\sqrt{\dfrac{2}{\omega_0^2 m}\left(E - \dfrac{p_0^2}{2m}\right)}$, $C_2 = \dfrac{p_0}{\omega_0 m}$;

(c) The two phases are related as $\varphi_2 = \pi - \varphi_1$, $\varphi_1 = \arctan\dfrac{C_2}{C_1}$.

10. A particle of mass m moves without friction along the parabola $y = kx^2$, where the y axis is pointed upwards (in the opposite direction to the weight). Find the frequency of small oscillations around the equilibrium position.

Solution. The expressions for the kinetic and potential energies are

$$K = (1 + 4k^2x^2)\frac{m\dot{x}^2}{2} \quad \text{and} \quad U = mgkx^2.$$

The linearization close to the equilibrium position, $x = 0$, is necessary only for the kinetic energy. As a result, we obtain $\omega_0^2 = 2gk$.

11. In the system illustrated in Fig. 7.5, the length of the spring itself is $b < a$. The parameter of elasticity of the spring is k. A particle of mass m moves along the horizontal line.

(a) Calculate the frequency of small oscillations around the equilibrium position. There is no friction.
(b) Check that in the case $b = a$ small oscillations will not be harmonic.
(c) Repeat the program of part (a) for the case $b > a$.

Answers:

$$(a) \qquad f = \frac{1}{2\pi}\sqrt{\frac{k}{m} \cdot \frac{a-b}{a}}.$$

(b) For the case $a = b$, the first relevant term of the potential is quartic, because there is no quadratic term. This implies that the oscillations are not harmonic ones.
(c) The potential energy is (here x is the coordinate of the mass m on the horizontal line)

$$U(x) = \frac{k}{2}(a^2 + b^2) + \frac{kx^2}{2} - kb\sqrt{a^2 + x^2}.$$

The plot of this potential has a typical form of a "Mexican hat", as shown in Fig. 7.6. In this case there are two distinct positions of equilibrium, one with $x_0 > 0$ and another one with $x_0 < 0$. The frequency of small oscillations near these positions of equilibrium is the same in both cases,

$$f = \frac{1}{2\pi}\sqrt{\frac{k}{m} \cdot \frac{b^2 - a^2}{a^2}}.$$

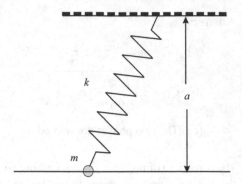

Fig. 7.5 Illustration for
Exercise 11

One can see that the situation with these two solutions is such that the potential energy satisfies the symmetry condition $U(-x) = U(x)$, however the solutions corresponding to small oscillations near equilibrium do not satisfy this symmetry. In other words, the system has to "choose" to stay in the $x_0 > 0$ or $x_0 < 0$ region. In the modern physics such situations are called Spontaneous Symmetry Breaking (SSB) and represent a very important element of our general understanding of modern theories. The theories with SSB are common in different branches of Modern Physics. In particular, the SSB is likely responsible for the existing masses of all elementary particles and, finally, for the masses of all bodies and media which we deal with.

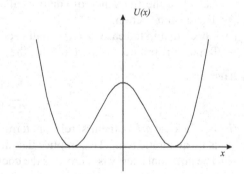

Fig. 7.6 "Mexican hat" – type
potential, typical for theories
with Spontaneous symmetry
breaking

12. As shown in Fig. 7.7, a system consists of a wire of fixed length, l, and a spring with elasticity k and unstretched length l. The distance between the points A and B is also given by l. Assume that the angle φ_0 corresponds to the equilibrium position, where the potential energy has minimum. Find the frequency of small oscillations around this equilibrium position.

Solution. Suppose that the spring reaches the size $L > l$, corresponding to φ_0 such that the system is in the equilibrium. Let us calculate the potential energy function of a small angle deviation from equilibrium, δ.

The spring deformation, Δx, is given by

$$\Delta x = l\left(2\sin\frac{\varphi_0}{2} - 1\right) + l\delta\cos\frac{\varphi_0}{2} - \frac{l\delta^2}{4}\sin\frac{\varphi_0}{2} + O(\delta^3),$$

and the gravitational potential energy (relative to the equilibrium position) is

$$U_{\text{grav}} = -mgl\left[\sin(\varphi_0 + \delta) - \sin\varphi_0\right] \approx mgl\left[\frac{\delta^2\sin\varphi_0}{2} - \delta\cos\varphi_0\right].$$

Together with the expression for elastic potential energy, $k\Delta x^2/2$, one can write the potential energy $U(\varphi_0 + \delta)$ as a function of δ and use

$$\omega_0^2 = \frac{1}{ml^2}\frac{d^2U}{d\varphi^2}\bigg|_{\varphi_0}.$$

Finally, we get

$$\omega_0^2 = \frac{g}{2l}\sin\varphi_0 + \frac{k}{m}\left[\frac{1}{2}\cos\varphi_0 + \frac{1}{4}\sin\frac{\varphi_0}{2}\right].$$

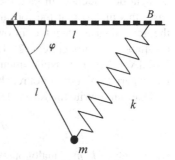

Fig. 7.7 Illustration of
Exercise 3

13. Determine the frequencies of small oscillations around the equilibrium position of the following potentials (in all cases, U_0 and L are constant):

(a) $U(x) = \frac{U_0}{2}\cos\frac{x}{L}$, (b) $U = U_0\cosh^2\frac{x}{L}$,

(c) $U = U_0(-\frac{x^2}{2L^2} + \frac{x^4}{4L^4})$, (d) $U = e^{-\frac{x^2}{2L^2}} \cdot \frac{4\sqrt{e}U_0}{3L}x$.

In all cases above sketch the potential showing the point of minimum (or one of them if there are several points of minimum).

Answer: $\omega_0^2 = \frac{2U_0}{mL^2}$ in all cases.

14. For all harmonic oscillators of the previous problem, determine explicit solutions for $q(t) = x(t) - x_0$, where x_0 is a local position of minimum of the potential for the cases:

(a) $q(0) = q_0$, $\dot{q}(0) = 0$;   (b) $q(0) = -q_0$, $\dot{q}(0) = v_0$.

Answer: (a) $q(t) = q_0 \cos \omega_0 t$,   (b) $q(t) = -q_0 \cos \omega_0 t + \dfrac{v_0}{\omega_0} \sin \omega_0 t$.

7.4 Damped Oscillations

Let us consider a more complicated case when, in addition to the restoring force, proportional to $q(t)$, there is an additional resistance (dissipative) force proportional to velocity,

$$F = m\ddot{q} = -\alpha\dot{q} - kq. \tag{7.23}$$

Note that the form of the friction force considered above is not the most general possible. We know that the frictional force between a body and a flat surface is not dependent on the magnitude of velocity, it depends only on the normal force. The study of this case can be done separately, as proposed in one of the exercises in this section. On the other hand, the damping effects, such as resistance of air or another fluid, are dependent on the absolute value of the velocity, but it is not necessarily a linear dependence. In fact, the linear form is valid only for slow motion. In our case, this means that the formula (7.23) can be seen as a good approximation of the general case when the following conditions are satisfied: (1) The frequency of oscillations is moderate; (2) The resistance force depends on velocity; (3) The initial velocity and displacement are sufficiently small. In this case the velocity of the material point does not reach high values and thus the Eq. (7.23) is justified.

One can rewrite the Eq. (7.23) in the form

$$\ddot{q} + 2\lambda\dot{q} + \omega_0^2 q = 0, \tag{7.24}$$

where $\omega_0^2 = k/m$ (analogously to the harmonic oscillator, discussed previously) and the new quantity $\lambda = \alpha/(2m)$ is called the *damping coefficient*.

To solve Eq. (7.24), we will use the same approach which was applied before. It is important to remember that behind our *ad hoc* calculations there is a rigorous mathematical formulation, in particular, the theorem of existence of the fundamental system of solutions of the homogeneous linear equation. This theorem simplifies our work considerably, because all that we need are two linearly independent particular solutions of (7.24). Assuming

$$q = e^{\beta t},$$

we find the following characteristic equation for the coefficient β:

$$\beta^2 + 2\lambda\beta + \omega_0^2 = 0.$$

The solution of this equation has the form

$$\beta = -\lambda \pm \sqrt{\lambda^2 - \omega_0^2} = -\lambda \pm i\sqrt{\omega_0^2 - \lambda^2}.$$

One can distinguish between the three cases, namely

1) **Case of high viscosity.** The damping coefficient is large, with $\lambda^2 > \omega_0^2$. In this case,

$$\beta_{1,2} = -\lambda \pm \sqrt{\lambda^2 - \omega_0^2} < 0,$$

and the general solution of (7.23) has the form

$$q(t) = C_1 e^{\beta_1 t} + C_2 e^{\beta_2 t}, \qquad (7.25)$$

with β_1 and β_2 defined above. The result indicates that there are actually no oscillations. After a sufficiently long time the system approaches the equilibrium position exponentially. This behavior is typical for a system with high resistance and relatively small elasticity.

An important observation is that the solution (7.25) does not necessarily have a qualitatively simple behavior. Depending on the initial conditions, an oscillator of this type can show a great increase of the coordinate and then return or not to the equilibrium position. In the case when in the given physical problem it is interesting to study the solution for time intervals smaller than any of the two numbers β_1^{-1} and β_2^{-1}, the motion to the equilibrium position may not be observed.

2) **Marginal case**, when $\lambda = \omega_0$. In this case the solution is given by

$$q(t) = e^{-\lambda t}(C_1 t + C_2). \qquad (7.26)$$

Qualitatively, there is not much difference with the previous case. For time periods much greater than λ^{-1}, the system will approach the equilibrium position exponentially.

3) **Oscillations with small damping**, corresponding to $\lambda^2 < \omega_0^2$. In this case,

$$\sqrt{\lambda^2 - \omega_0^2} = i\omega,$$

where $\omega = \sqrt{\omega_0^2 - \lambda^2} > 0$ is a real number. The general solution of the dynamical equation has the following complex form:

$$q(t) = e^{-\lambda t}\left(C_1 e^{i\omega} + C_2 e^{-i\omega}\right).$$

Correspondingly, a general real solution is

$$q(t) = e^{-\lambda t}[C_1 \cos \omega t + C_2 \sin \omega t] = A_0 e^{-\lambda t} \cos(\omega t + \varphi), \qquad (7.27)$$

where $A_0 = \sqrt{C_1^2 + C_2^2}$ is called the initial amplitude of oscillation. For small damping, the result (7.27) can be seen as the description of a harmonic oscillation with variable amplitude,

$$A(t) = A_0 e^{-\lambda t}.$$

The constants A_0 and φ can be related to the initial values q_0 and $v_0 = \dot{q}(t_0)$, in a way similar to the case of a harmonic oscillator. We leave this calculation as an exercise.

We can calculate the time rate of change for the total mechanical energy of the system. It is obvious that mechanical energy is not conserved because of the damping, which is responsible for the dissipation of energy. The calculation in the general case is quite extensive, although being simple. For this reason, we leave the general case as an exercise for an interested reader, and consider only the particular case with $\lambda \ll \omega_0$. This case has an advantage of admitting technically simple solutions, because all the terms proportional to λ are small and can be disregarded.

The time dependence of the potential energy can be written as

$$U(t) = \frac{kq^2}{2} = \frac{kA_0^2}{2} e^{-2\lambda t} \cos^2(\omega t + \varphi).$$

Furthermore, disregarding the terms proportional to λ, we obtain, by direct calculation, the expression for the kinetic energy,

$$K(t) \cong \frac{mA_0^2 \omega^2}{2} e^{-2\lambda t} \sin^2(\omega t + \varphi).$$

Note that in general $\omega^2 \neq k/m$. However, in the given approximation we can indeed set $\omega \approx \omega_0$. In this way we arrive at the following expression for the mechanical energy:

$$E(t) \cong \frac{mA_0^2}{2} \omega^2 e^{-2\lambda t} = E_0 e^{-2\lambda t}, \tag{7.28}$$

where E_0 is the mechanical energy at the initial time instant. Thus, one can see that the time dependence of the mechanical energy has an exponential form, i.e., it is proportional to $A^2(t)$. In many interesting examples of the macroscopic "harmonic" oscillations, the energy dependence has a behavior which is approximately described by (7.28).

Exercises

1. For the case $\lambda^2 < \omega_0^2$, find relations between the sets $\{q_0, v_0\}$ and $\{A_0, \varphi\}$.

Answer: $q_0 = A_0 \cos \varphi, \quad v_0 = -A_0(\lambda \cos \varphi + \omega \sin \varphi)$.

2. Calculate an exact formula for $E(t)$ and compare with Eq. (7.28).

Answer:

$$E(t) = \frac{mA_0^2\omega_0^2}{2}e^{-2\lambda t}\left[1 + \frac{m\lambda\sqrt{\omega_0^2 - \lambda^2}}{\omega_0^2}\sin 2(\omega t + \varphi) + \frac{\lambda^2}{\omega_0^2}\cos 2(\omega t + \varphi)\right].$$

3. At a planet X, the astronauts decided to measure the acceleration of local gravity, g_X. For this end they decided to use a mathematical pendulum of length $L = 20$ cm. The initial amplitude was given by $A_0 = 8°$, and the measurements indicated that the amplitudes A_n, after n oscillations, are $A_{10} = 7.36°$, $A_{20} = 6.77°$ and $A_{30} = 6.23°$. The total time measured after 30 periods of oscillation was 18.27 s. Using these data, find g_X and λ.

Answer: $g_X = 21.3$ m/s^2, $\lambda = 0.0137$ Hz.

4. Consider a damped oscillator system with $\lambda = \omega_0$. Set the initial conditions such that $q(0) = 0$ and that during the interval of time between $t = 0$ and $T = 1/\omega_0$, the value of q increases such that it reaches A. At which moment of time will the coordinate q reach the value $0.01\,eA$?

Solution. Using the initial data, we find the particular solution of interest, $q(t) = v_0 t\,e^{-\lambda t}$. It is easy to see that $q(t)$ reaches A at the instant $t = 1/\lambda$, if the initial velocity is $v_0 = e\lambda A$. For the instant of time $t = x/\lambda$ which is necessary to reach $0.01\,eA$, we find the equation $xe^{-x} = 0.01$, which can be solved only approximately, by numerical methods. The answer is $t \approx 6.47/\lambda$.

Observation. We advise the reader to plot the function $q(t)$.

5. For the system from the previous exercise, verify whether there is an initial condition for which the coordinate $q(t)$ changes sign only once during its motion. Discuss whether it is possible to find two sign changes.

Answer: Using the solution (7.26) we can easily see that the necessary condition for $q(t) = 0$ with $t > 0$ is $C_1 \cdot C_2 < 0$. Two sign changes are obviously impossible.

6. Repeat the considerations of the previous exercise, in the case of a high viscosity, given by $\lambda = 2\omega_0$.

7.5 Forced Oscillations

Consider now the more general case with the presence of the damping force and also an external force $f(t)$. The equation of motion becomes more complicated,

$$\ddot{q} + 2\lambda\dot{q} + \omega_0^2 q = \frac{f(t)}{m}. \tag{7.29}$$

The term on the right side is called the driving term, or just the source term. When this term is zero, we have a homogeneous linear equation. Otherwise, the equation is non-homogeneous. To explore the solution of this equation, it is necessary to use two theorems about differential equations. The reader may find the proofs of these theorems in books on differential equations, e.g., in [4,7].

Theorem 2. *The general solution of the linear non-homogeneous equation is given by the sum of the general solution of the homogeneous equation and a particular solution of the non-homogeneous equation.*

Theorem 3. *If the free term of the non-homogeneous linear equation is given by the sum of several functions,*

$$f(t) = f_1(t) + f_2(t) + \ldots + f_N(t),$$

then the particular solution of the equation has the form

$$q(t) = q_1(t) + q_2(t) + \ldots + q_N(t),$$

where, for any $i = 1, 2, \ldots, N$, the expression $q_i(t)$ is a particular solution of the equation with the unique source $f_i(t)$.

For our case, Theorem 1 means that the general solution of (7.29) is given by the sum of the general solution of the homogeneous equation, (7.27), and a particular solution of the non-homogeneous equation (7.29). This particular solution is fixed in the sense that there is no dependence on arbitrary constants of integration. In other words, all the necessary constants of integration are included in the general solution of the homogeneous equation.

To simplify the consideration, we will study only the case with small viscosity, $\lambda < \omega_0$. As an example, we find the solution to a particular case in which

$$f(t) = f_0 \cos \gamma t, \tag{7.30}$$

where $\gamma \neq 0$. It is important to mention that the solution of the equation with the source (7.30) can be helpful for the case of an arbitrary physically relevant kind of sources. In order to understand this, let us remember Theorem 2. The explanation is that a wide class of functions, $f(t)$ can be expanded into Fourier series, and then the solution of equation (7.29) can be always found in the form of a series of solutions for sources of the form (7.30).

Let us remember that according to Theorem 1, our purpose is to find a particular solution, no matter which one.[4] One can look for the particular solution for the source (7.30) in the form

$$q(t) = B_1 \cos \gamma t + B_2 \sin \gamma t, \tag{7.31}$$

---

[4] It is easy to see that the difference between two solutions of the non-homogeneous equation is a solution of the homogeneous equation. In other words, the difference between two such solutions can always be compensated by adjusting the constants of integration and eventually disappear at the moment when we implement initial conditions.

where B_1 and B_2 are some constants that we have to define by replacing into the equation. It is worth warning that B_1 and B_2 do not represent arbitrary constants of integration and defining them does not involve the use of initial data. Taking derivatives of (7.31), we find

$$\dot{q}(t) = -B_1\gamma\sin\gamma t + B_2\gamma\cos\gamma t,$$
$$\ddot{q}(t) = -B_1\gamma^2\cos\gamma t - B_2\gamma^2\sin\gamma t.$$

Substituting these formulas in the Eq. (7.29), we find

$$\cos\gamma t\left(-B_1\gamma^2 - 2\lambda B_2\gamma + B_1\omega_0^2\right)$$
$$+ \sin\gamma t\left(-B_2\gamma^2 + 2\lambda\gamma B_1 + B_2\omega_0^2\right) = \frac{f_0}{m}\cos\gamma t.$$

Comparing the two sides of the last equation and requesting the identity for (separately) the terms proportional to $\cos\gamma t$ and $\sin\gamma t$, we obtain the equations for B_1 and B_2 namely

$$B_2(\omega_0^2 - \gamma^2) + 2\lambda\gamma B_1 = 0,$$
$$B_1(\omega_0^2 - \gamma^2) - 2\lambda\gamma B_2 = \frac{f_0}{m}. \tag{7.32}$$

Solving the last system of equations, we arrive at

$$B_2 = -\frac{2\lambda\gamma}{\omega_0^2 - \gamma^2}B_1, \qquad B_1 = \frac{f_0(\omega_0^2 - \gamma^2)}{m\left[(\omega_0^2 - \gamma^2)^2 + 4\lambda^2\gamma^2\right]}. \tag{7.33}$$

Finally, the general solution for the forced oscillator with the source (7.30) can be written as follows:

$$q(t) = A_0 e^{-\lambda t}\cos(\omega t + \varphi) + B_1\cos\gamma t + B_2\sin\gamma t, \tag{7.34}$$

where B_1 and B_2 are defined by (7.33). According to this solution, the damped system will "forget" the initial data after a sufficiently long period of time, namely, after the time $\Delta t \gg 1/\lambda$. Asymptotically, after a long time, the solution depends only on the external force.

It is especially interesting to consider the case when the frequency of the source (7.30) is close to the proper frequency of the free oscillator. Consider $\gamma = \omega_0 + \varepsilon$, where $|\varepsilon| \ll \omega_0$ is a small deviation of the frequency of the source in relation to the frequency of the oscillator itself. In this case it is sufficient to account for the first-order terms in ε. In this way we obtain the relation

$$\gamma^2 - \omega_0^2 = \omega_0^2 + 2\omega_0\varepsilon - \omega_0^2 = 2\omega_0\varepsilon.$$

For the sake of simplicity, we can also consider λ to be very small (low damping case). Then, one can determine the amplitude of stationary forced oscillations (the word "stationary" means that after a long period of time, the first term on the right side of the Eq. (7.34) disappears) as

$$B = (B_1^2 + B_2^2)^{\frac{1}{2}} = B_1 \left[1 + \frac{4\lambda^2 \gamma^2}{(\omega_0^2 - \gamma^2)^2} \right]^{1/2} \approx B_1 \left[1 + \frac{\lambda^2}{\varepsilon^2} \right]^{1/2},$$

where

$$B_1 \approx \frac{f_0}{m} \cdot \frac{2\varepsilon \omega_0}{4\omega_0^2 \varepsilon^2 + 4\lambda^2 \omega_0^2 + 8\lambda^2 \omega_0 \varepsilon}$$

$$= \frac{\varepsilon f_0}{m \left(2\omega_0 \varepsilon^2 + 2\omega_0 \lambda^2 + 4\lambda^2 \varepsilon \right)}. \tag{7.35}$$

The first observation is that for $\lambda \ll \varepsilon$ we have

$$B_1 \approx \frac{f_0}{2m\omega_0} \cdot \frac{1}{\varepsilon},$$

i.e., the amplitude increases considerably in the vicinity of the value $\varepsilon = 0$. This phenomenon is called *resonance*. The study of resonance behavior is very important, for it has several physical consequences and applications in science and technology. If friction is not negligible, the amplitude of the oscillations depend on the relationship between ε and λ.

Exercises

1. Analyze the behavior of the system

$$\ddot{x} + 2\lambda \dot{x} + \omega_0^2 x = f \sin(\alpha - \omega_0 t)$$

with initial conditions $x(0) = 0$ and $\dot{x}(0) = 0$. Consider $\lambda \ll \omega_0$.

Answer: In the first approximation in the ratio λ/ω_0, the solution is given by

$$x = B \cos \omega_0 (t - t_0) - Be^{-\lambda t} \cos(\omega t - \omega t_0 + 2\omega_0 t_0),$$

$$\omega^2 = \omega_0^2 - \lambda^2, \qquad B = \frac{f}{2\lambda \omega}. \tag{7.36}$$

2. Analyze the behavior of the system

$$\ddot{x} + 2\lambda \dot{x} + \omega_0^2 x = f \sin \frac{\omega_0 t}{10}$$

with initial conditions $x(0) = \dot{x}(0) = 0$. Assume $\lambda \ll \frac{\omega_0}{10}$. What is the qualitative difference if compared to the previous problem?

Hint. Use Eqs. (7.32) and (7.34) to analyze the difference.

3. Solve the same problem as in the previous exercise, but this time with a source given by $f(t) = f_1 \sin(10\omega_0 t) + f_2 \sin(\omega_0 t / 10)$.

4. A block of mass M is stuck at the end of a spring with constant k, placed in the horizontal. The block can oscillate horizontally on a rough surface, with coefficient of friction given by μ. The initial position of the block is such that the spring is not deformed, and its initial velocity is v_0. Assuming a very small friction μ, calculate the number of cycles n that the system performs until it stops.

Observation. This problem has relatively a high level of difficulty.

Answer: $n = \frac{v_0}{4\mu g}\sqrt{\frac{k}{M}}$.

5. A body of mass m_1 is connected to another body, with mass m_2, through a spring with constant k. The system is placed on a horizontal surface without friction. There is no rotation. Calculate the period of small fluctuations around the equilibrium position.

Answer: $T = 2\pi\sqrt{\mu/k}$, where μ is the reduced mass of the system, $\mu = \frac{m_1 m_2}{m_1 + m_2}$.

7.6 Multidimensional Oscillators

Until now we discussed only the case of one-dimensional movement, and correspondingly one-dimensional oscillators. In fact, many problems in physics and related areas require the study of motion in systems with many degrees of freedom. In many situations, harmonic oscillations in systems with many degrees of freedom play a specially important role. For example, this is the case for molecules with more than one atom, solid bodies, nanotubes and other compounds. In all these cases the study of classical oscillatory behavior is a necessary element before quantum aspects of the theory can be explored. Because of the introductory character of the present course, we will not address the issue of multidimensional oscillators in full detail. Instead, in the last section of this chapter our intention is to explain the main ideas of multidimensional oscillators in the simplest possible way.

Multidimensional oscillators appear when particle can move in different directions or when a system of particles or a body can perform movements, which are more complicated than the one-dimensional ones. An obvious example of a two-dimensional oscillator is the mathematical pendulum, that can move in two angular directions. A good example of a three-dimensional oscillator is the same pendulum, when we take into account the possibility of small radial oscillations of the wire. The latter case is somehow typical, because here different degrees of freedom have different geometric nature (two angles and lengths) and consequently different dimensions. Yet, both cases can be studied analytically. We leave to the reader to work out these two examples as exercises.

   In summary, the solution of the harmonic oscillator is done by means of the special procedure of reduction of a coupled multidimensional system to a set of independent one-dimensional oscillators, called *normal modes* of the initial system. The transition to normal modes represents a linear change of variables. As far as the harmonic oscillator is a kind of universal construction, which does not depend much on the physical origin of the problem, the same reduction procedure can be used in very different areas of Physics, for example, in the study of the spectra of large molecules and in defining the masses of elementary particles. The approach is always the same. Starting from the very beginning, the process includes the search for the equilibrium position, then derivation of equations for small oscillations around the equilibrium position and, finally, reduction to normal modes. In analytical mechanics, there are regular methods to set up the equations of motion for complicated systems and also more powerful methods for solving equations for the linear systems (see, e.g., [11,21]). As we already mentioned above, here we intend to treat only relatively simple cases.

   Mathematically, the search for normal modes is possible in most cases when there is no friction. When the damping terms are present, the procedure typically does not work and one has to deal with a system of several coupled differential equations. The introduction of sources (external forces) can be done in the same manner as in the case of one-dimensional oscillator, so we are not going to discuss this issue here. In order to illustrate the general statements done above, we consider two simple examples of two-dimensional oscillators and leave a few other cases as exercises.

Example 4. Consider the motion of a particle with mass m in a field of a force with the potential

$$U(x,y) = \frac{k_1}{2} x^2 + \frac{k_2}{2} y^2 , \qquad (7.37)$$

where $k_{1,2}$ are distinct elasticity coefficients. Let us write the full expression for the mechanical energy of the system,

$$E = K + U = \text{const}, \qquad K = \frac{m}{2} \dot{x}^2 + \frac{m}{2} \dot{y}^2 . \qquad (7.38)$$

We note that in both kinetic and potential energy parts the terms corresponding to the directions x and y are separated. This means that they already represent the normal modes of the two-dimensional oscillator. There is no exchange of energy between the sectors of $x(t)$ and $y(t)$, therefore we can assume that the energy in each sector is conserved separately.[5]

   Assuming that the energy for x and y modes is independently conserved, the equations for $x(t)$ and $y(t)$ can be easily calculated by taking time derivatives (similarly to how it was done in Chap. 6). We obtain

---

[5] There is a mathematically rigorous way to prove this statement in Lagrangian (Analytical) Mechanics. Here we rely on our intuition.

$$m\ddot{x}^2 + k_1 x = 0, \qquad m\ddot{y}^2 + k_2 y = 0, \qquad (7.39)$$

with the general solutions in the form

$$x(t) = x_0 \cos(\omega_1 t + \varphi_1), \qquad y(t) = y_0 \cos(\omega_2 t + \varphi_2),$$

$$\text{where} \qquad \omega_{1/2}^2 = \frac{k_{1/2}}{m}. \qquad (7.40)$$

Here x_0, y_0 and φ_1, φ_2 are amplitudes and initial phases of the normal modes. Obviously, the motions along the axes OX and OY occur independently. At the same time, if some special relation between the frequencies takes place, one can observe certain coherence between the solutions for two oscillators. In order to study this possibility, we have to write the expressions for the two periods of oscillations,

$$T_1 = \frac{2\pi}{\omega_1} = 2\pi\sqrt{\frac{m}{k_1}}, \qquad T_2 = \frac{2\pi}{\omega_2} = 2\pi\sqrt{\frac{m}{k_2}}. \qquad (7.41)$$

In the case $T_1/T_2 = n_1/n_2$, where n_1 and n_2 are natural numbers, the two-dimensional motion is periodic. This criterion is equivalent to the following relation:

$$k_1/k_2 = (n_2/n_1)^2. \qquad (7.42)$$

In the case of periodic motions, all possible particle orbits are closed, i.e., the particle will always return to a starting position with exactly the same velocity. In contrast, in the case where the condition (7.42) is not satisfied, the particle never passes twice through exactly the same position on the plane XOY, despite after some time passing very close to this position. The meaning of "very close" is that after a finite time interval, the particle will arrive at a distance smaller than ε from its initial position, for any $\varepsilon > 0$.

A very special case is when $k_1 = k_2$. Then, among the possible motions of the oscillator, one can mention, for example, the elliptical motion in the plane XOY and one-dimensional (linear) oscillation in the same plane.

Example 5. Let us now discuss another particular case, when the initial coordinates do not represent the normal modes of the oscillator. We consider a concrete illustration, in order to avoid an unnecessary complexity.

A particle of mass m moves in the potential

$$U(x,y) = \frac{k}{2}(x^2 + 2xy + 2y^2). \qquad (7.43)$$

In this case, the kinetic energy is still given by the formula (7.38). Thus, the main difference is the presence of the mixed term in the expression for the potential energy, (7.43). It should be noted that in many problems with multidimensional harmonic oscillator, terms are mixed also in the expression for kinetic energy. We choose a simplified version here only for the purpose of displaying the most basic example.

The idea of the search for the normal mode is to perform such a change of variables that does not break the linearity of the equations of motion and, at the same time, eliminates mixing terms. Let us perform a rotation on the plane XOY, by an arbitrary angle, α. The new variables are $Q_1(t)$ and $Q_2(t)$,

$$x = Q_1 \cos \alpha + Q_2 \sin \alpha,$$
$$y = -Q_1 \sin \alpha + Q_2 \cos \alpha. \tag{7.44}$$

Given the transformation (7.44), the form of the kinetic energy term does not change, such that

$$E = K + U = \text{const}, \qquad K = \frac{m}{2} \dot{Q}_1^2 + \frac{m}{2} \dot{Q}_2^2. \tag{7.45}$$

The new expression for potential energy is given by

$$
\begin{aligned}
U(Q_1, Q_2) = \frac{k}{2} \Big[&Q_1^2 (\cos^2 \alpha + 2 \sin^2 \alpha - 2 \cos \alpha \sin \alpha) \\
&+ Q_2^2 (2 \cos^2 \alpha + \sin^2 \alpha + 2 \cos \alpha \sin \alpha) \\
&+ 2 Q_1 Q_2 (\cos^2 \alpha - \cos \alpha \sin \alpha - \sin^2 \alpha) \Big].
\end{aligned}
\tag{7.46}
$$

Solving the equation

$$\cos^2 \alpha - \cos \alpha \sin \alpha - \sin^2 \alpha = 0$$

to remove the mixed term, we find

$$\sin^2 \alpha = \frac{5 - \sqrt{5}}{10}, \qquad \cos^2 \alpha = \frac{5 + \sqrt{5}}{10}. \tag{7.47}$$

A very simple calculation shows that using the new variables, the potential energy (7.43) is reduced to the sum of the two independent oscillator terms

$$U(Q_1, Q_2) = \frac{k_1}{2} Q_1^2 + \frac{k_2}{2} Q_1^2, \tag{7.48}$$

where

$$k_1 = \frac{3 - \sqrt{5}}{10} k \qquad \text{and} \qquad k_2 = \frac{3 + \sqrt{5}}{10} k.$$

As we have already mentioned, the kinetic energy has the standard form (7.45) and therefore the system already belongs in the class considered in Example 1. The transition to the original variables can be accomplished by the transformation inverse to (7.44),

$$Q_1 = x \cos \alpha - y \sin \alpha,$$
$$Q_2 = x \sin \alpha + y \cos \alpha. \tag{7.49}$$

Obviously, for the given particular example, coefficients k_1 and k_2 do not satisfy the criterion (7.42), therefore the motion is not periodic and the trajectory is not closed.

Exercises

1. Consider in greater detail the special case of Example 1, when $k_1 = k_2 = k$. Classify the possible shapes of closed trajectories. For this purpose establish the set of possible initial data for which the trajectory of motion is given by the following curves:

(a) A circle of radius R.

(b) A straight line segment of length L. Review the possible dependencies on the time in this case.

(c) An ellipse with semi-axes a and b. Show that the first two trajectories represent special cases of this one.

(d) A closed curve with the aspect of the number 8.

Answers and solution:

$$(a) \quad x(0) = R\cos\varphi_0, \qquad y(0) = R\sin\varphi_0,$$
$$\dot{x}(0) = \mp\omega R\sin\varphi_0, \quad \dot{y}(0) = \pm\omega R\cos\varphi_0.$$

$$(b) \quad x(t_0) = 0, \qquad y(t_0) = 0, \quad \dot{x}(t_0) = v_0\cos\varphi_0,$$
$$\dot{y}(t_0) = v_0\sin\varphi_0, \quad \text{where} \quad v_0^2 = \frac{kL^2}{4m}.$$

$$(c) \quad x(0) = a\cos\varphi_0\cos\chi_0 - b\sin\varphi_0\sin\chi_0,$$
$$y(0) = a\cos\varphi_0\sin\chi_0 + b\sin\varphi_0\cos\chi_0,$$
$$\dot{x}(0) = \mp\omega(a\sin\varphi_0\cos\chi_0 + b\cos\varphi_0\sin\chi_0),$$
$$\dot{y}(0) = \mp\omega(a\sin\varphi_0\sin\chi_0 - b\cos\varphi_0\cos\chi_0).$$

In the case (c), the angle χ_0 indicates the slope of the semi-major axis of the ellipse with respect to axis OX.

To understand that the solution described in the point (d) is impossible, one has to note that its existence contradicts the uniqueness of the solution of the Cauchy problem for the oscillator due to the fact that the particle must pass the point $r = 0$ with the same velocity in both cases (b) and (d).

2. Consider the motion of a mathematical pendulum with two degrees of freedom. Study the case of small oscillations and show that the problem reduces to the situation described in Example 1 with $k_1 = k_2 = k$. Calculate the frequency in terms of the length of the wire L, the mass of the pendulum m, and the acceleration of gravity g. Explain why in the first approximation it is not necessary to consider the component $\dot{z}$ of the speed. After making explicit calculations, try to use arguments based on symmetry and dimensions to obtain the frequency.

Answer: In this case the expression $I = mL^2$ has a role of mass (I is called the moment of inertia and will be discussed in the next chapter). The result is $\omega^2 = g/L$.

3. Consider the movement of the spherical pendulum which is almost mathematical. Such a pendulum has three degrees of freedom, two of them angular as in the previous Exercise and, on the top of that, the wire has finite elasticity coefficient κ. Consider small oscillations of this system. Compare with the result of Exercise 2.

Hint and answer. To solve this problem, just write the expressions for the kinetic and potential energy when the spring length takes the value $L + \rho(t)$, where $|\rho(t)| \ll L$. It turns out, that in the linear approximation the degree of freedom related to the deformation of the spring is independent of the angular degrees of freedom and so the problem is reduced to the previous case, with $\omega_1^2 = g/L$ and $\omega_2^2 = \kappa/m + g/L$.

4. In the case considered in Example 2, sketch the lines of the normal zero modes $Q_1 = 0$ and $Q_2 = 0$ in the plane XOY.

5. Modify the formula for the coefficients of the potential energy, Eq. (7.43), such that after the transition to the normal modes, Q_1, Q_2, the frequencies of the new system satisfy the condition (7.42) with n_1/n_2 of your choice.

Hint. Try to solve this problem without many calculations, which is perfectly possible.

Chapter 8
Dynamics of Rotational Movements

Abstract In this chapter we discuss the dynamics of rotational movements. The main object of study is the rotational dynamics of a solid body, but we will also consider the rotational movement of a particle or a system of particles. As we shall see in what follows, the dynamics of rotational motion has many analogies with the dynamics of the translational motion. At the same time, there are important differences, in part because of the cyclic nature of the angular variable. This property means that after increasing the value of the variable for a certain period (equal to 2π in case of an angle), the body comes back to the same position. We will formulate the laws of rotational dynamics and find the corresponding conservation law.

8.1 Torque and Moment of Inertia

We already know that the kinematical description of rotational movements of a body requires an introduction of the angular velocity vector, $\vec{\omega}$, and the angular acceleration, $\vec{\beta} = d\vec{\omega}/dt$. Both are vector quantities and the direction of $\vec{\omega}$ indicates the orientation of the instantaneous axis of rotation.

Our first task will be to construct a suitable formulation of angular dynamics. One might expect that this dynamics would be similar or even equivalent, at least to some extent, to Newton's laws for translational movements that we have studied in the previous chapters. For translational motion, we know that Newton's Second Law for a particle of mass m is written as $\mathbf{F} = m\mathbf{a}$, where $\mathbf{F}$ is the force exerted on the particle and $\mathbf{a}$ is its acceleration.

It is natural to assume that the dynamics of rotations is also described by a similar equation, say

$$\vec{\tau} = I\vec{\beta}, \tag{8.1}$$

I.L. Shapiro and G. de Berredo-Peixoto, *Lecture Notes on Newtonian Mechanics*,
Undergraduate Lecture Notes in Physics, DOI 10.1007/978-1-4614-7825-6_8,
© Springer Science+Business Media, LLC 2013

where $\vec{\tau}$ is a quantity that contains the information of the interaction, and I is an analogue of the particle mass, namely some intrinsic quantity related to the inertia of rotational movements.

It is possible to deduce a law similar to Newton's Second Law and therefore find the meaning of I and $\vec{\tau}$. Let us consider the equation of Newton for a single particle and multiply both sides vectorially by the position vector of the same particle. Thus, we obtain

$$\mathbf{F} = m\frac{d\mathbf{v}}{dt} \quad \Longrightarrow \quad \mathbf{r} \times \mathbf{F} = \mathbf{r} \times m\frac{d\mathbf{v}}{dt} = \frac{d}{dt}(\mathbf{r} \times \mathbf{p}),$$

where we take into account that $\mathbf{p} = m\mathbf{v}$ and

$$\frac{d\mathbf{r}}{dt} \times \mathbf{v} = \mathbf{v} \times \mathbf{v} \equiv 0.$$

We can express this result as

$$\vec{\tau} = \frac{d\mathbf{L}}{dt}, \tag{8.2}$$

where

$$\vec{\tau} = \mathbf{r} \times \mathbf{F} \tag{8.3}$$

is called *torque*, or moment of force $\mathbf{F}$ acting on the particle and

$$\mathbf{L} = \mathbf{r} \times \mathbf{p} \tag{8.4}$$

is called *angular momentum* vector of the particle.

The Eq. (8.2) is obtained from Newton's Second Law. In this equation, when compared to the original form of Newton's Second Law, $\mathbf{F} = \dot{\mathbf{p}}$, the force is replaced by the torque and the linear momentum is replaced by the angular momentum. As we will discover later on, Eq. (8.2) is exactly the desired law (8.1), but for now the relationship between torque and angular momentum from one side and rotation from another side does not look clear. In Analytical Mechanics, this relationship appears naturally, but in this book we will use the formalism which is much more based on physical intuition. Therefore, our understanding of the law (8.1) will be improved by consideration of particular examples of mechanical systems.

To better understand the relation between the torque $\vec{\tau}$ and rotation, consider the system illustrated in Fig. 8.1. The axis of rotation of the rigid bar is fixed orthogonally to the plane of the page, then the angular acceleration vector has fixed direction and its direction will be clockwise or counterclockwise. That means that the vector $\vec{\beta}$ also has direction which is fixed by the axis of rotation. The direct verification shows that the same applies to the torque $\vec{\tau}$.

Let us check that the torque, $\vec{\tau}$, is related to the angular acceleration. The two forces, $\mathbf{F}_1$ and $\mathbf{F}_2$, mutually parallel and orthogonal to the bar, are applied to the same bar. The points of application are at different distances, l_1 and l_2, of the axis of

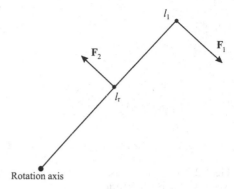

Fig. 8.1 Rotation around a
fixed axis perpendicular to the
plane of the figure

rotation. Experience shows that when the directions of the forces are opposite, the
equilibrium condition of the bar has the form

$$F_1 l_1 = F_2 l_2. \tag{8.5}$$

This empirical relation is called *law of the lever*. That means that, different from the
translational movements, the force has no independent relevance for the rotational
movement. According to hypothesis of (8.1) and (8.3), and to Eq. (8.5), each of the
forces, F_1 and F_2, contributes to the angular acceleration and this contribution is
in the form of the product of the force F with the arm l, i.e., $F_1 l_1$ and $F_2 l_2$. When
these quantities are identical in magnitude and act in opposite directions, the system has no angular acceleration, according to (8.5). The same force, according to
formula (8.5), can produce greater or smaller angular acceleration, depending on
the value of the arm. In the case of equilibrium, the products $F_1 l_1$ and $F_2 l_2$ have the
same value and obviously contribute to angular accelerations with the same magnitude and in opposite directions, so that the resulting angular acceleration is zero.
Compared to Eq. (8.3) for torque, it is easy to see that the vectors $\vec{\tau}_1$ and $\vec{\tau}_2$,
produced by the forces F_1 and F_2, have opposite directions.

Now let us consider the situation which is a little bit more complicated, as shown
in Fig. 8.2. As before, the axis of rotation is fixed.

If the bar with length $r = |\mathbf{r}|$ is rigid, the force can be decomposed into compo-
nents which are parallel and perpendicular to the bar. The parallel component always
cancels with the tension of the bar itself. Therefore, only the orthogonal component
of the force F_1 is relevant for angular acceleration, and the equilibrium condition is
given by

$$F_1 \cdot r \cdot \sin \alpha = F_2 r.$$

So, the force F contributes to the angular acceleration through the expression
$N = F \cdot r \cdot \sin \alpha$, where $r = |\mathbf{r}|$. It is clear that this term is nothing else but the magnitude of the torque, defined in (8.3), for the force acting on a single particle. Our
consideration shows that the notion of torque can be applied to describe the rotational movements of dynamic systems of many particles, including rigid bodies.

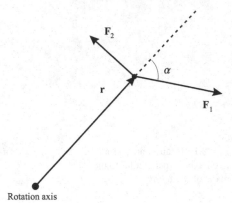

Fig. 8.2 More general case, where a force is not perpendicular to **r**

The general definition of torque in Classical Mechanics is the following. As we have explained before, any body can be seen as a set of point particles. Considering a system of N particles, the position of the i-th particle relative to the origin of the coordinate system O, is specified by the radius-vector $\mathbf{r}_i$. If such a particle is subject to the force $\mathbf{F}_i$, the total torque on the system, with respect to the point O, is defined as the vector sum of the torques of each force,

$$\vec{\tau} = \sum_{i=1}^{N} \mathbf{r}_i \times \mathbf{F}_i. \tag{8.6}$$

Note that the torque vector depends on the choice of a point O. This is different from the case of a total force, where the last is the same in any inertial reference frame.

The next step is to understand the rotational analogue of Newton's Second Law in the form of the Eq. (8.1), that is to identify the nature of the quantity I. The dimensionality of this quantity can be obtained from the relation (8.1). As far as the angle in radians has no dimension, we can write

$$[\beta] = \frac{1}{[T]^2} \qquad \text{and also} \qquad [\tau] = [F] \cdot [r] = \frac{[M] \cdot [L]^2}{[T]^2}. \tag{8.7}$$

Here the brackets indicate the dimension of the quantity of our interest. The symbols $[M]$, $[L]$ and $[T]$ denote the fundamental dimensions: mass, length and time, respectively. Thus, the meaning of $[\beta] = 1/[T]^2$ is that β has dimension inverse to the square of time.

It is clear from (8.1) and (8.7) that I is a scalar quantity, with dimensionality given by

$$[I] = [M][L]^2, \tag{8.8}$$

equivalent, in the International System of Units (IS), to $[I] = \text{kg} \cdot \text{m}^2$. In the Gaussian system, the dimension of moment of inertia is given by $[I] = \text{g} \cdot \text{cm}^2$.

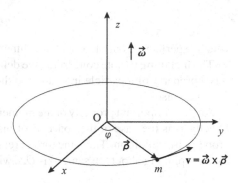

Fig. 8.3 Particle spinning
around a fixed axis

In fact, one can use the formula (8.8) as a hint to define I, that we will call *moment of inertia*. However, before discussing further details of this concept, it is worth making an important observation.

Comparing the formulas (8.2) and (8.1), we can note that what we seek is the proportionality between the angular momentum and the angular velocity of the body. This proportionality is valid for a particle and for bodies of some special form, but it is *not* correct in the general case, when a rigid body has an arbitrary distribution of mass, i.e., does not have axial symmetry around the axis of rotation. In the general case, the vectors $\vec{\omega}$ and **L** are not parallel and the rotation requires higher level of mathematical description than the one adopted in this book. In particular, the moment of inertia in the general case is a tensor quantity. An introduction to tensor calculus is not a difficult study (see, for example, [2, 28, 29]), but the use of tensors is out of the scope of the present textbook, which is designed as elementary introduction. The reader can study more general formulations of rotational dynamics in the books on Analytical Mechanics, e.g., [10, 11, 20, 21]. In this book, we will not consider the general case, hence our approach is, generally, restricted to the case of a fixed axis of rotation and to bodies with axial symmetry around the axis of rotation. In other words, we study mainly those cases where $\vec{\omega}$ and **L** are parallel vectors and the tensor nature of the moment of inertia is not relevant.

Let us now obtain an explicit formula for the moment of inertia I of a rotating body. Consider the simplest case, namely a point-like particle with mass m rotating around the axis OZ, as shown in Fig. 8.3, with angular velocity $\dot{\varphi} = \omega$. One can write the expression for the angular momentum as

$$\mathbf{L} = \vec{\rho} \times \mathbf{p} = m\rho v \hat{\mathbf{k}} = m\rho^2 \omega \hat{\mathbf{k}} = m\rho^2 \vec{\omega}.$$

Using this expression, Eq. (8.2) can be written, in this particular case, as

$$\vec{\tau} = \frac{d\mathbf{L}}{dt} = m\rho^2 \frac{d\omega}{dt} \hat{\mathbf{k}} = m\rho^2 \vec{\beta},$$

which can be identified with Eq. (8.1). This identification can be accomplished if we define the moment of inertia I as follows:

$$I = m\rho^2, \tag{8.9}$$

which is perfectly consistent with the dimension $[M][L]^2$.

The first thing to note concerning the definition of momenta of inertia (8.9) is that it is a property of a particle in relation to the given axis. If the axis is not specified, I has no sense.

Another important property of the moment of inertia and torque is their additivity. Exactly as is the case of the concepts of mass and force, we need to introduce this property by definition. Thus, the torque related to the origin of a system of N particles all spinning about the same axis OZ, with the same angular velocity is given by

$$\vec{\tau} = \sum_{i=1}^{N} \mathbf{r}_i \times \mathbf{F}_i = \vec{\beta} I, \qquad \text{where} \qquad I = \sum_{i=1}^{N} I_i \tag{8.10}$$

and $I_i = m_i \rho_i^2$ is the moment of inertia of the i-th particle with respect to the axis of rotation and ρ_i is the distance between the i-th particle and the same axis.

The formula (8.10) describes the rotational dynamics of a system of particles or a body, that rotates around a fixed axis. It is easy to see that this equation is exactly the realization of the Eq. (8.1), which was the main target of our consideration.

In order to illustrate how the formalism of rotational dynamics works, let us consider a few simple examples.

Example 1. A point-like body of mass M is connected to the axis Oz through a lightweight bar with length r, perpendicular to this axis. The force F acting on the body is perpendicular to the axis and to the bar as shown in Fig. 8.4. Calculate the angular acceleration of the body.

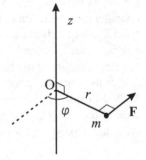

Fig. 8.4 Rotational motion of a particle subject to a force or a torque (Example 1)

Solution. One can use Newton's Second Law. The linear acceleration is given by $a = F/m$ and using the relation between a and angular acceleration β, we arrive at

$$\beta = \frac{a}{r} = \frac{F}{Mr}.$$

Solution. One can use the rotational analogue of Newton's Second Law, (8.1). The torque acting on the body is a vector whose modulus is given by $\tau = Fr$ and the moment of inertia can be written as $I = Mr^2$. Then we write

$$\beta = \frac{\tau}{I} = \frac{Fr}{Mr^2} = \frac{F}{Mr},$$

i.e., the result is the same as in the first case.

It is important to realize that the new formalism, based on formula (8.1), allows us to solve many problems that would be extremely difficult or impossible to deal with using only the Second Law of Newton for translational motion. Typically, most of these problems are about the motion of a body or a system of bodies, which consist of an infinite number of points. In order to elaborate the necessary tools to solve these problems, we compute the moments of inertia of some extended bodies. After that, the reader is supposed to gain sufficient experience to calculate the moments of inertia for other bodies of interest.

The idea of our approach is as follows: we divide an extensive body (which is supposed to be continuous) into small portions, such that each one of them can be treated as a massive point. As it was mentioned before, the number of such points tends to infinity when the dimension of each of them becomes close to zero (continuous limit). At the same time, the moment of inertia can be expressed formally as a sum of the moment of inertia of point-like particles. It should be remembered that in this procedure it is understood that the moment of inertia is an additive quantity. In practice, we apply a continuous approach and simply take an appropriate integral.

Example 2. A homogeneous bar has mass M and length L. Calculate the moment of inertia of the bar in relation to an axis passing through one of its ends, which is perpendicular to the bar.

Solution. The contribution of the segment of the bar between r and $r + dr$ can be obtained by noting that the mass contained in this portion is given by

$$dM = \frac{M\,dr}{L}.$$

Then, the moment of inertia of this infinitesimal part will be

$$dI(r) = r^2 dM = \frac{r^2 M\,dr}{L}.$$

By integrating throughout the bar, from 0 to L, we find the total moment of inertia of the bar,

$$I = \int_0^L \frac{dI}{dr}\,dr = \frac{1}{L}\int_0^L Mr^2\,dr = \frac{ML^2}{3}. \qquad (8.11)$$

The result can be easily generalized to the case when

(a) The axis goes through a different point, and

(b) The angle between the bar and the axis is different from 90°. We leave these generalizations as exercises.

Example 3. A homogeneous disk has radius R and mass M. Calculate the moment of inertia for the axis that is perpendicular to its plane and pass through its center.

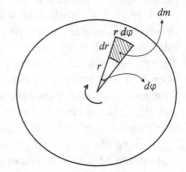

Fig. 8.5 Illustration for calculating the moment of inertia of a homogeneous disc relative to the axis perpendicular to the disk that passes through the origin (Example 3)

Solution. We will use polar coordinates, r and φ, as shown on the illustration in Fig. 8.5. The element of area is given by $rdr \cdot d\varphi$. Since the density of the disk is constant, we obtain the mass of the element of the area in the form

$$dm = \frac{M}{\pi R^2} \cdot rdrd\varphi,$$

and the differential element of the moment of inertia can be written as

$$dI = r^2 dm = \frac{M}{\pi R^2} r^3 dr d\varphi.$$

By integrating over the entire disk, we obtain

$$I = \int_0^{2\pi} d\varphi \int_0^R \frac{M}{\pi R^2} r^3 dr = \frac{2\pi}{\pi} \cdot \frac{M}{R^2} \cdot \frac{R^4}{4} = \frac{MR^2}{2}. \tag{8.12}$$

Example 4. Consider the same disk of the previous example, but this time the axis of rotation lies in the plane of the disc and pass through its center. Calculate the moment of inertia.

Solution. In this case, according to Fig. 8.6, we can write

$$dI = r^2 \sin^2 \varphi \, dm = \frac{M}{\pi R^2} \sin^2 \varphi d\varphi \cdot r^3 dr.$$

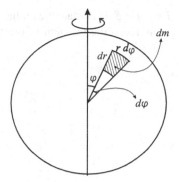

Fig. 8.6 Illustration for the calculation of the moment of inertia of a homogeneous disc in relation to an axis belonging to the plane of the disc (Example 4)

After integration, we obtain

$$I = \frac{M}{\pi R^2} \int_0^{2\pi} \sin^2 \varphi \, d\varphi \int_0^R r^3 dr = \frac{M}{\pi R^2} \cdot \pi \cdot \frac{R^4}{4} = \frac{MR^2}{4},$$

where we have used

$$\int_0^{2\pi} \sin^2 \varphi \, d\varphi = \frac{1}{2} \int_0^{2\pi} (1 - \cos 2\varphi) \, d\varphi = \pi.$$

Solution. It is possible to reach the same answer using the result of Example 1 considered above, and symmetry-based arguments. In the previous case,

$$I_z = \int r^2 dm,$$

where the axis OZ is orthogonal to the plane of the disc. In terms of the Cartesian coordinates on the plane of the disk, x and y, the I_z can be expressed as

$$I_z = \int (x^2 + y^2) \, dm = \int x^2 dm + \int y^2 dm.$$

Note that each of the integrals on the right hand side of the equation above is exactly the moment of inertia relative to the axis belonging to the plane of the disc and passing through its center. This is exactly the moment of inertia which we intend to derive in this example. Moreover, among the possible axes belonging to the plane of the disc and passing through the origin, there are no privileged axis. This is due to the axial symmetry of the disc, namely the disc does not change after being rotated by some angle φ around the axis OZ. As a consequence, the moment of inertia for the axis OX, $\int y^2 dm$, is equal to the moment of inertia for the axis OY, $\int x^2 dm$. So, one can write

$$I_x = \int x^2 dm = \frac{I_z}{2} = \frac{MR^2}{4}.$$

This alternative solution is an example of how symmetry considerations can help to solve problems in Mechanics.

Example 5. Calculate the moment of inertia of a spherical shell of radius r and mass m and also the moment of inertia of a solid sphere of radius R and mass M (See Fig. 8.7).

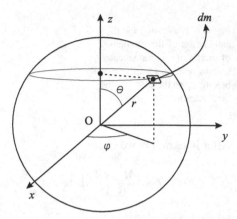

Fig. 8.7 Illustration for calculating the moment of inertia of a spherical shell (Example 5)

Solution. First we compute the moment of inertia of the shell. This calculation can be performed in different ways. Here we shall use spherical coordinates, i.e., angular coordinates φ and θ. The differential elements of mass and moment of inertia are given by

$$dm = \frac{m}{4\pi r^2} r^2 \, d\varphi \sin\theta d\theta,$$

$$dI = r^2 \sin^2\theta dm = \frac{mr^2}{4\pi} d\varphi \sin^3\theta d\theta.$$

Integrating over the angles, we obtain

$$I_{shell} = \frac{mr^2}{4\pi} \int_0^{2\pi} d\varphi \int_0^{\pi} \sin^3\theta d\theta \tag{8.13}$$

$$= -\frac{mr^2}{2} \int_0^{\pi} \left(1 - \cos^2\theta\right) d\cos\theta = \frac{2mr^2}{3}.$$

The calculation for the case of a solid sphere can be done independently, or using the result (8.12) for the disc, or using the result (8.13) for the spherical shell. Using the latter approach, we can divide the spherical body of radius R into thin shells of radius r and infinitesimal thickness dr. The mass of each shell is equal to the volume, $dV = 4\pi r^2 dr$, multiplied by the density of the solid sphere, given by $\rho = M \cdot \left(4\pi R^3/3\right)^{-1}$. In order to use (8.13), we just need to trade m to $dm = \rho dV$, according to

$$dI = \frac{2r^2 dm}{3} = \frac{2r^2}{3} 4\pi r^2 \frac{3M}{4\pi R^3} dr = \frac{2r^4 M}{R^3} dr.$$

Integrating over r, we arrive at the result

$$I_{shere} = \frac{2M}{R^3} \int_0^R r^4 dr = \frac{2MR^2}{5}. \tag{8.14}$$

A very interesting question is how to evaluate the moment of inertia of a body about a given axis, α_1, if we know the moment of inertia of the same body in relation to another axis, α_2. For the sake of simplicity, consider the case in which the axes are parallel. To understand the importance of this issue, just try to calculate, for example, the moment of inertia of a sphere in respect of any axis that does not pass through its center. The calculation becomes very cumbersome. Thus it would be desirable to have a relationship between the moments of inertia. And then, instead of performing new complicated calculations, in case of the sphere one can just use the result (8.14).

The general problem is resolved by the Parallel Axes Theorem, or Steiner (or Huygens – Steiner) Theorem:

Theorem 1. *Consider a two-dimensional body of mass M, with center of mass at point O, and the axis α_0 perpendicular to the plane of the body and passing through the point O. Suppose that the moment of inertia of the body in relation to α_0 is given by I_0. Then the moment of inertia in relation to another axis, α, parallel to α_0 and lying on the distance R from it, is given by*[1]

$$I = I_0 + MR^2. \tag{8.15}$$

Proof. To show this important statement, we can use an approach which can be called "discretization". This means we replace the body as a collection of particles. Let us consider a system with N particles with masses $m_1, m_2, \ldots, m_N$. The i-th particle ($i = 1, 2, \ldots, N$) has radius-vector $\mathbf{r}_i^0$ in relation to the center O, placed on the axis α_0 and the radius vector $\mathbf{r}_i$ in the reference frame with the center O_α, placed on the axis α. It is easy to prove that the choice of the point O_α on the axis α is not relevant.

It will be useful to introduce the unit vector $\hat{\mathbf{n}}$, parallel to the axes α and α_0. The first important observation is that the contribution of a particle to the moment of inertia I depends on the modulus of the vector product

$$\rho_i = |\hat{\mathbf{n}} \times \mathbf{r}_i| = r_i \sin \theta_i,$$

where θ_i is the angle between the vectors $\hat{\mathbf{n}}$ and $\mathbf{r}_i$. In other words, ρ_i is the radius in cylindrical coordinates of the point m_i, in the coordinate system with the axis α, or simply the projection of $\mathbf{r}_i$ on the plane perpendicular to axis. One can write

$$I = \sum_{i=1}^N m_i \rho_i^2. \tag{8.16}$$

---

[1] The theorem works also for three-dimensional bodies with the property that each of its plane sections orthogonal to α_0 has the center of mass belonging to α_0. A more general formulation is possible in Analytical Mechanics [11, 21].

Similarly, we write

$$I_0 = \sum_{i=1}^{N} m_i \left(\rho_i^0 \right)^2 , \tag{8.17}$$

where $\rho_i^0 = \left| \hat{\mathbf{n}} \times \mathbf{r}_i^0 \right|$. Without loss of generality, we assume that the points O_α and O belong to the same plane perpendicular to the axes α and α_0.

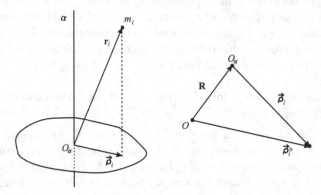

Fig. 8.8 The illustrations on the *left* and *right* show respectively the relation between the vectors $\mathbf{r}_i$ and ρ_i, and the relationship between the vectors ρ_i and ρ_i^0. In the *right*, both axes α and α_0 are perpendicular to the plane

To demonstrate the relationship (8.15), one can introduce the vectors $\vec{\rho}$ and $\vec{\rho}_i^0$ in the plane perpendicular to the axes α and α_0. In this case, $\rho_i = |\vec{\rho}_i|$ and also $\rho_i^0 = |\vec{\rho}_i^0|$. Moreover,

$$\vec{\rho}_i = \vec{\rho}_i^0 - \mathbf{R},$$

where $\mathbf{R} = \overrightarrow{OO_\alpha}$ (see Fig. 8.8).

Now we can write the expression for the moment of inertia, I, as

$$I = \sum_{i=1}^{N} m_i \rho_i^2 = \sum_{i=1}^{N} m_i \vec{\rho}_i \cdot \vec{\rho}_i = \sum_{i=1}^{N} m_i \left(\vec{\rho}_i^0 - \mathbf{R} \right) \cdot \left(\vec{\rho}_i^0 - \mathbf{R} \right)$$

$$= \sum_{i=1}^{N} m_i \left(\vec{\rho}_i^0 \cdot \vec{\rho}_i^0 - 2\mathbf{R} \cdot \vec{\rho}_i^0 + \mathbf{R} \cdot \mathbf{R} \right)$$

$$= \sum_{i=1}^{N} m_i \left(\rho_i^0 \right)^2 - 2\mathbf{R} \cdot \sum_{i=1}^{N} m_i \vec{\rho}_i^0 + MR^2 , \tag{8.18}$$

where we used the notation $M = \sum_i m_i$. It is easy to see that the second term of the last expression is zero because the axis α_0 passes through the center of mass of the system of particles. Furthermore, the first term is exactly (8.17), that is the moment of inertia I_0. Thus, we have demonstrated the relation (8.15). $\square$

Let us return to the description of rotational dynamics. We can solve some interesting problems, such as the problem of physical pendulum, which, unlike the mathematical pendulum, has an arbitrary mass distribution. Let us consider a body which is hung in a point which is different from its center of mass. The body performs small oscillations in a fixed plane around the equilibrium position.

To gain experience, we will first use the Eq. (8.1) to analyze the movement of the mathematical pendulum (see Fig. 8.9). This problem has been addressed in Chap. 7, but here we will use Eq. (8.1) instead of the Second Law of Newton. The origin of the polar system of coordinates can be placed at the point where the wire is hung. In this case, $\beta = \ddot{\varphi}$, where $\varphi = \varphi(t)$ is the angle of deviation from the equilibrium position, measured positive in the counterclockwise direction. Of course, the tension of the wire, $\mathbf{T}$, produces no torque, because $\mathbf{T} \parallel \mathbf{r}$. The angle between $\mathbf{r}$ and the gravity acceleration, $\mathbf{g}$, is given by φ. Then we obtain the following expression for the magnitude of the torque:

$$\tau = |\vec{\tau}| = mg \cdot |\mathbf{r}| \cdot \sin \varphi = mgL \sin \varphi.$$

Then we have

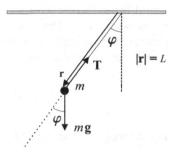

Fig. 8.9 Analysis of the mathematical pendulum

$$\beta - \ddot{\varphi} = -\frac{\tau}{I} = -\frac{mgL \sin \varphi}{mL^2} - -\frac{g}{L} \sin \varphi, \qquad (8.19)$$

i.e., the same result as in Chap. 7. The choice of the negative sign is because the torque $\vec{\tau}$ has an opposite direction to the vector of the angle $\vec{\varphi}$. Assuming a small angle we obtain $\ddot{\varphi} = -g\varphi/L$, such that the frequency of the oscillations is given by

$$\omega_0 = \sqrt{\frac{g}{L}} \quad \text{and the period,} \quad T = 2\pi \sqrt{\frac{L}{g}}.$$

Now consider an example of a physical pendulum, consisting from a homogeneous bar of mass M and length L hanging on its end. The situation can be illustrated by Fig. 8.9 replacing a mathematical pendulum by a physical pendulum.

Before solving this problem, it is necessary to address the following question: In the case of a homogeneous and uniform gravitational field, $\mathbf{g}$, each element of mass has a weight. Can we replace all these particular gravitational forces by a single total weight applied at some point, such that the torque would be the same? Is this hypothesis true? If so, what is the point at which we can apply the total weight?

To answer these questions, one has to remember that the two possible important dynamical effects of a force are the acceleration of the body and its accelerated rotation (in case of non-zero torque). Due to the additive property of forces, a system of forces can be replaced by a resultant force producing the same acceleration. Similar to this, one can expect that a system of forces acting on a rigid body is dynamically equivalent to the unique resultant force applied at some point. This point should be chosen so that the resultant force produces the same torque relative to the origin compared to the set of real forces acting on the parts of the body.

The hypothesis described above is correct in the case of homogeneous gravity, and we can show it by demonstrating the following statement about the torque:

Theorem 2. *Consider a body or a system of particles, with a total mass M, subject to the action of a homogeneous gravitational field, g. The weight force and its torque with respect to an arbitrary point O are the same as the weight and torque applied to the center of mass point of the body.*

Proof. We shall use the same approach of discretization which was already proved useful before. Consider a set of points with masses m_i, $i = 1,2,\ldots,N$, and with position vectors r_i. Our task is to show that the weight and its torque are the same compared to the case when the same gravitational field g acts on a single particle with the mass and radius-vector, respectively,

$$M = \sum_i^N m_i \quad \text{and} \quad R = \frac{1}{M} \sum_i^N m_i r_i.$$

For the weight force, we find

$$P = \sum_i^N m_i g = g \sum_i^N m_i = Mg, \tag{8.20}$$

i.e., we could easily show the desired equivalence. For the case of torque, we can make a similar transformation,

$$\vec{\tau} = \sum_i^N r_i \times (m_i g) = \left\{ \sum_i^N r_i m_i \right\} \times g = MR \times g. \tag{8.21}$$

The last formula shows the equivalence for the torque, with respect to any starting point O.

Thus, the force produced by the uniform gravitational field can be always applied to the center of mass of the body. For a homogeneous bar, the center of mass is $L/2$ from its end. Coming back to the physical pendulum problem, we can write, for the absolute value of torque,

$$\tau = \frac{MgL\sin\varphi}{2},$$

where L is the length of pendulum. If the moment of inertia of the body for the same axis is given by I, we obtain

$$\ddot{\varphi} = -\frac{\tau}{I} = -\frac{MgL}{2I}\sin\varphi.$$

For small angles, $\sin\varphi \approx \varphi$ and the equation above reduces to the equation for the harmonic oscillator,

$$\ddot{\varphi} = -\frac{\tau}{I} = -\frac{MgL}{2I}\varphi,$$

describing oscillations with the frequency given by

$$\omega_0 = \sqrt{\frac{MgL}{2I}}.$$

For a homogeneous bar, hanged by one of its ends, we obtain

$$\ddot{\varphi} = -\frac{\tau}{I} = -\frac{MgL/2}{ML^2/3}\sin\varphi = -\frac{3}{2}\frac{g}{L}\sin\varphi.$$

For small angles, we have $\ddot{\varphi} = -\frac{3g}{2L}\varphi$, and thus the frequency of the oscillations is

$$\omega_0 = \sqrt{\frac{3g}{2L}}.$$

This result is approximately 1.23 times greater than the frequency of a mathematical pendulum of the same length and mass.

Exercises

1. Calculate the moment of inertia of a thin homogeneous bar of mass M and length L, with respect to the axis which makes an angle α with the bar and passes by the point of the bar that is at a distance x from one of its ends.

Answer:

$$I = \frac{M\sin^2\alpha}{3}\left[x^3 + (L-x)^3\right].$$

2. Use the result for the moment of inertia of the homogeneous disk, from Example 3, to recalculate the moment of inertia of the massive sphere. Then calculate the moment of inertia of the spherical shell using only the result for the solid sphere.

Hint: In the last case, consider dI corresponding to dr.

3. Calculate the moment of inertia relative to the axis $y = z = L/\sqrt{2}$ of a solid homogeneous cylinder of mass M, with the shape defined by the relations

$$0 \le x \le H \qquad y^2 + z^2 \le L^2 .$$

Hint. Use the Steiner theorem.

Answer:

$$I = \frac{3}{2} ML^2 .$$

4. Verify the theorem of Steiner for the particular case of the bar of Example 2, but this time with different choice of axis. In particular, consider

(a) Axis passing through the center of mass of the bar and being perpendicular to the bar;

(b) Axis perpendicular to the bar and passing on the distance $a < L/2$ from its center of mass. Try to solve the problem of calculating the moments of inertia with and without use of integrals.

5. Check the sign of the formula (8.19) using the vector product $\vec{\tau} = \mathbf{r} \times \mathbf{F}$.

6. A heavy homogeneous disk of mass M and radius R can rotate without friction around the vertical axis OZ, as illustrated in Fig. 8.10. Each of the two light propellants produces a force of magnitude F. Calculate the angular acceleration and velocity due to both propellants after the time t from the beginning of the movement. How would the result change if each of the propellants had mass m?

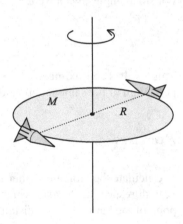

Fig. 8.10 Exercise 6

Answer:

$$\beta = \frac{4F}{(4m + M)R}, \qquad v = \beta Rt.$$

7. A homogeneous disk of mass M and radius R is hanging at a point of its edge and performs small oscillations in the plane perpendicular to the plane of the disc. Calculate the frequency of these oscillations. How would the result change if small oscillations were in the plane of the disc?

Answers:

$$\omega_\perp = \sqrt{\frac{4g}{5R}}, \qquad \omega_\| = \sqrt{\frac{2g}{3R}} .$$

8.2 Angular Momentum of a System of Particles

Consider a particle with radius-vector $\mathbf{r}$ and momentum $\mathbf{p} = m\mathbf{v}$. In the previous section, the angular momentum of this particle with respect to the origin has been defined as $\mathbf{L} = \mathbf{r} \times \mathbf{p}$, as a rotational analogue of the momentum (in a similar fashion torque is the rotational analogue of force). Using the formula for the vector product (see Mathematical Appendix), we can write

$$\mathbf{r} \times \mathbf{p} = \begin{vmatrix} \hat{\mathbf{i}} & \hat{\mathbf{j}} & \hat{\mathbf{k}} \\ x & y & z \\ p_x & p_y & p_z \end{vmatrix} \tag{8.22}$$

$$= \hat{\mathbf{i}} \left(y p_z - z p_y \right) + \hat{\mathbf{j}} \left(z p_x - x p_z \right) + \hat{\mathbf{k}} \left(x p_y - y p_x \right).$$

Therefore, the vector components of $\mathbf{L}$ can be written as

$$L_x = y p_z - z p_y, \quad L_y = z p_x - x p_z, \quad L_z = x p_y - y p_x, \tag{8.23}$$

and its absolute value is equal to

$$L = \sqrt{L_x^2 + L_y^2 + L_z^2} = r p \sin \alpha,$$

where α is the angle between the vectors $\mathbf{r}$ and $\mathbf{p}$. An important property of the angular momentum is that it is an additive quantity, i.e., the angular momentum of a system of particles is equal to the sum of the angular momentum of each individual particle. This means that a system of N particles with the radius vectors $\mathbf{r}_i$ and linear momentum $\mathbf{p}_i = m_i \mathbf{v}_i$ has total angular momentum given by

$$\mathbf{L} = \sum_{i=1}^{N} \mathbf{L}_i = \sum_{i=1}^{N} \mathbf{r}_i \times \mathbf{p}_i. \tag{8.24}$$

For a better understanding of these general statements, let us consider two simple examples.

Example 6. Consider a particle of mass m moving in the plane XOY. The vector $\mathbf{L}$ is parallel to the axis OZ. The direction and the modulus of this vector depend on the vectors $\mathbf{r}$ and $\mathbf{p} = m\mathbf{v}$ of the particle.

Example 7. A body with cylindrical (axial) symmetry that rotates around its axis of symmetry, OZ, has angular momentum parallel to the axis OZ. It is easy to see that this is a direct consequence of axial symmetry. Suppose the opposite, i.e., there is a component of angular momentum which is orthogonal to OZ. Then the rotational movement of the body breaks down the axial symmetry, hence we meet a contradictory situation. The reader can also verify this result by means of calculations in different examples. It is also possible to prove a general statement, which we leave as an exercise.

To calculate the angular momentum of a body, we use the "discretization" approach, which was also adopted in the previous section for the proof of the theorem of Steiner. Let us consider a system of N particles, of masses $m_1, m_2, \ldots, m_N$. The i-th particle (here $i = 1, 2, \ldots, N$) has radius-vector $\mathbf{r}_i$ and the linear momentum $\mathbf{p}_i = m_i \mathbf{v}_i$. The total angular momentum of the system is given by

$$\mathbf{L} = \sum_{i=1}^{N} \mathbf{r}_i \times \mathbf{p}_i = \sum_{i=1}^{N} m_i \mathbf{r}_i \times \mathbf{v}_i, \tag{8.25}$$

We can express $\mathbf{r}_i$ in the coordinate system related to the center of mass of the system of particles. For this end, assume that the position of the center of mass in the original reference frame is described by the vector $\mathbf{R}$. As it is shown in Fig. 8.11, $\mathbf{r}_i = \mathbf{R} + \mathbf{r}'_i$, where $\mathbf{r}'_i$ is the position vector of i-th particle relative to the center of mass. Therefore,

$$\mathbf{v}_i = \frac{d\mathbf{r}_i}{dt} = \frac{d\mathbf{R}}{dt} + \frac{d\mathbf{r}'_i}{dt} = \mathbf{V} + \mathbf{v}'_i,$$

where $\mathbf{V}$ is the velocity of the center of mass.

Thus, we can rewrite (8.25), using $\sum_i m_i = M$, as

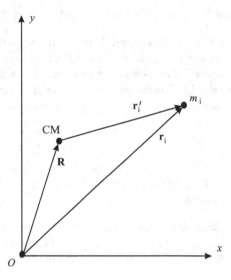

Fig. 8.11 Position of the i-th particle relative to the reference of the center of mass

$$\mathbf{L} = M\mathbf{R} \times \mathbf{V} + \mathbf{R} \times \left(\sum_{i=1}^{N} m_i \mathbf{v}'_i \right) + \left(\sum_{i=1}^{N} m_i \mathbf{r}'_i \right) \times \mathbf{V} + \sum_{i=1}^{N} \mathbf{r}'_i \times m_i \mathbf{v}'_i.$$

One can note that the velocity of the center of mass in relation to reference frame (C) is always equal to zero, also the position of the center of mass in the same reference frame is also identically zero. Therefore, we obtain

$$\sum m_i \mathbf{v}_i' = 0 \qquad \text{and} \qquad \sum m_i \mathbf{r}_i' = 0,$$

and the final expression is given by

$$\mathbf{L} = M\mathbf{R} \times \mathbf{V} + \mathbf{L}' \qquad (8.26)$$

where $\mathbf{L}'$ is the angular momentum of the system of particles in relation to its center of mass. The last formula shows how to transform the angular momentum related to the reference frame (C) into any other reference frame. This transformation is similar to the one for the linear momentum of a system of particles, $\mathbf{P} = \mathbf{P}' + M\mathbf{V}$. The main difference between these two cases is that we definitely have $\mathbf{P}' \equiv 0$ in the reference frame of the center of mass (C). Obviously, this is not the case for angular momentum, because the body or system of particles may be rotating and consequently $\mathbf{L}'$ is not necessarily zero in the reference frame (C).

The calculation of the angular momentum of a system of particles with respect to its center of mass is, generally, a problem that requires a considerable effort. However, we can consider some specific examples in which this calculation is easier. A useful simplification is to treat a system of particles as a rigid body.

Consider a rigid body rotating about an axis OZ. The angular momentum with respect to the origin (belonging to the axis OZ) is given by the integral on the volume (V) of the body,

$$\mathbf{L} = \iiint\limits_{(V)} \mathbf{r} \times \mathbf{v}\, dm = \iiint\limits_{(V)} \mathbf{r} \times (\vec{\omega} \times \mathbf{r})\, dm, \qquad (8.27)$$

where $dm \equiv \rho\, dV$, where ρ is the density of the mass of the body and dV the differential element of the volume. The angular velocity of rotation of the body is given by $\vec{\omega} = \omega\hat{\mathbf{k}}$, that is the vector parallel to the axis OZ and pointing out to the positive direction of z. Using the identity (see Mathematical Appendix)

$$\mathbf{r} \times (\vec{\omega} \times \mathbf{r}) = \vec{\omega}\, r^2 - \mathbf{r}(\vec{\omega} \cdot \mathbf{r}),$$

one can calculate

$$\mathbf{L} = \omega\hat{\mathbf{k}} \iiint\limits_{(V)} r^2 dm - \omega \iiint\limits_{(V)} \left(x\hat{\mathbf{i}} + y\hat{\mathbf{j}} + z\hat{\mathbf{k}} \right) z\, dm, \qquad (8.28)$$

where x, y and z are Cartesian components of the position vector, $\mathbf{r} = x\hat{\mathbf{i}} + y\hat{\mathbf{j}} + z\hat{\mathbf{k}}$ and the integral is calculated over the entire volume of the body. Recalling that

$$r^2 = x^2 + y^2 + z^2,$$

one can see that the term with z^2 in the first expression is compensated by the term with z^2 in the second one. Taking this into account, we can rewrite the Eq. (8.28) in the form

$$\mathbf{L} = \omega\hat{\mathbf{k}} \underset{(V)}{\iiint} \left(x^2 + y^2\right) dm - \omega \underset{(V)}{\iiint} \left(x\hat{\mathbf{i}} + y\hat{\mathbf{j}}\right) z\, dm. \qquad (8.29)$$

According to the last equation, the angular momentum of a rigid body, rotating with angular speed $\vec{\omega}$ around a fixed axis, may have all components non-zero, in the general case. However, as we have already mentioned above, the expression for the angular momentum can be simplified for some particular cases.

Suppose, for example, that a rigid body is homogeneous and has an axis of symmetry parallel to OZ. In this case, we can take the integral $\int_V xz\,dm$ in (8.29) and express it as follows:

$$\int_V xz\,dm = \rho \int z\,dz \int x\,dx\,dy = \rho \int z\,dz\, x_{CM} A(z),$$

where x_{CM}, y_{CM} and z_{CM} are coordinates of the center of mass, ρ is the density of mass of the rigid body, and $A(z)$ is the cross-sectional area of the rigid body at the level z. Since

$$\int \rho z A(z)\,dz = \int z\,dm = M z_{CM}$$

(remember $M = \int dm$), by applying the same consideration to the integral $\int_V yz\,dm$ we obtain

$$\mathbf{L} = I_z \omega\hat{\mathbf{k}} - M\omega z_{CM}\left(x_{CM}\hat{\mathbf{i}} + y_{CM}\hat{\mathbf{j}}\right), \qquad (8.30)$$

where $I_z = \int_V (x^2 + y^2)\,dm$ is the moment of inertia relative to the axis OZ. There is a direct relation between the last formula and the Eq. (8.26), we leave it as an exercise to the reader.

Finally, let us note that the expression (8.30) becomes even simpler if the center of mass of the rigid body has the coordinate $z = 0$. Then

$$\mathbf{L} = I_z \omega\hat{\mathbf{k}}. \qquad (8.31)$$

If the axis of rotation coincides with the axis of symmetry of the rigid body, then $x_{CM}\hat{\mathbf{i}} + y_{CM}\hat{\mathbf{j}} = 0$ and the angular momentum will be determined by the formula (8.31).

Another special case of the Eq. (8.30) is a flat body placed at the plane $z = 0$. If the axis of symmetry is parallel to OZ, the angular momentum $\mathbf{L}$ is given by the same formula, Eq. (8.31).

Exercises

1. Show that the angular momentum of a body which has axial symmetry around the axis OZ and is rotating around the same axis, has angular momentum in the direction of the axis OZ.

Hint. The easiest way is to use the discretization of the body. In this case, it is sufficient to take into account the contributions of two equal point-like masses with offsetting positions relative to the axis OZ.

2. Compare the formulas (8.26) and (8.30). Indeed, the latter is a particular case of the former. Identify the terms on the right side of the Eq. (8.30), in accordance with the general expression (8.26).

3. Calculate the angular momentum with respect to the origin of coordinate system for a particle of mass m, that performs a movement described by

$$z = h = \text{const.}, \quad x = R \cos \omega t, \quad y = R \sin \omega t.$$

Answer.

$$\mathbf{L} = -mhR\omega\left(\hat{\mathbf{i}}\cos \omega t + \hat{\mathbf{j}}\sin \omega t\right) + m\omega R^2 \hat{\mathbf{k}}.$$

8.3 Conservation of Angular Momentum

In previous sections we have learned that there is an analog for Newton's Second law, namely Eq. (8.1) for the dynamic of rotational motion. On the other hand, we know that Newton's Second Law applied to closed systems implies the law of conservation of momentum. The natural question is whether there is a similar law in the case of rotational movement, let us call it the law of conservation of angular momentum. The purpose of this section is to explore this conservation law.

Consider again the derivation of the Eq. (8.2) performed at the beginning of this chapter. When the resultant torque acting on a particle is zero, its angular momentum remains constant as a simple consequence of Newton's Second Law. At the same time, both Eqs. (8.2) and (8.1) correspond to the case of a single particle. Our purpose is to generalize it to a system of particles interacting among themselves through internal forces. It is important to remember that any macroscopic body can be seen as a system of particles, so that considering a system of N particles we arrive at the result that will have a general nature, including the case of a continuous material body.

Consider a closed system of N particles, i.e., a system which is free from an influence of external forces. According to Newton's Third Law, the internal forces can be grouped in pairs of action and reaction. Then Newton's Third Law requires that the force exerted by i-th particle over the j-th particle, $\mathbf{F}_{ij}$, satisfies the rule $\mathbf{F}_{ij} = -\mathbf{F}_{ji}$. What is the total torque produced by these two forces on the system of two particles? Assuming that the forces of action and reaction belong to the same line, for each pair one can conclude that the torque produced by them, with respect to any point of space, is identically zero. If one of the pairs does not satisfy this condition, the internal forces would produce an angular acceleration of the system, and this contradicts experimental results. Under these conditions, we verify that the overall torque produced by couple of forces is actually null.

Let us present a formal consideration of the arguments given above. For the two particles with labels 1 and 2, we obtain

$$\vec{\tau} = \mathbf{r}_1 \times \mathbf{F}_{21} + \mathbf{r}_2 \times \mathbf{F}_{12} = \mathbf{r}_1 \times \mathbf{F}_{21} - \mathbf{r}_2 \times \mathbf{F}_{21} = (\mathbf{r}_1 - \mathbf{r}_2) \times \mathbf{F}_{21}.$$

However, since for the couple of particles the unique special direction is the vector between them, we have

$$\mathbf{F}_{21} \parallel (\mathbf{r}_1 - \mathbf{r}_2), \tag{8.32}$$

such that $\vec{\tau} = 0$. Next, since all internal forces can be separated into pairs, the same relation (8.32) is true for all of them. Finally, in the absence of external forces, the total torque on a closed system of N particles is zero. This means that the angular momentum is an integral of motion, $\mathbf{L} = const$.

Let us now make an important observation. In the discussion presented above, we assumed that the forces of action and reaction do not only satisfy the third law of Newton $\mathbf{F}_{21} = -\mathbf{F}_{12}$, but also belong to the same line joining the two particles. This condition does not follow from Newton's third law and represents a qualitatively new implementation, which is indeed necessary for the description of dynamics of rotational movements. If the condition mentioned above was not necessarily satisfied, the total torque of these two particles would be non-zero and thus the law of conservation of angular momentum for a pair of particles would be violated.

It is obvious that the central forces, such as gravitational or electric, always act on the line joining the two particles. Any other option for the forces would lead to a direct contradiction to the isotropy of space. If we have two particles, there may be only one privileged direction, namely the one that links these particles. The situation is more delicate when considering magnetic forces, since they depend also on the velocities of the particles. In this case, the line joining the particles is not the unique privileged direction.[2] In the framework of Analytical Mechanics or in Field Theory one can obtain the conservation of angular momentum as a direct consequence of the isotropy of space. In this introductory course, we do not have the necessary tools to make this demonstration, so we assume that the mentioned condition is satisfied simply by definition.

Consider now the general case, when the external forces are present. For the i-th particle, we have

---

[2] The correct analysis of the magnetic forces can only be done in the scope of Relativistic Electrodynamics [17, 23]. This consideration shows a perfect agreement with the law of conservation of angular momentum.

$$\frac{d\mathbf{p}_i}{dt} = \mathbf{F}_i = \sum_{k \neq i} \mathbf{F}_{ki} + \mathbf{F}_i^{(ext)}.$$

In order to explore the dynamics of the rotational movement of the system of particles, one can take the derivative of angular momentum $d\mathbf{L}/dt$,

$$
\begin{aligned}
\frac{d\mathbf{L}}{dt} &= \sum_i \frac{d\mathbf{L}_i}{dt} = \sum_i \frac{d}{dt}(\mathbf{r}_i \times \mathbf{p}_i) = \sum_i \left(\frac{d\mathbf{r}_i}{dt} \times \mathbf{p}_i + \mathbf{r}_i \times \frac{d\mathbf{p}_i}{dt} \right) \\
&= \sum_i \left(\frac{1}{m} \mathbf{p}_i \times \mathbf{p}_i + \mathbf{r}_i \times \mathbf{F}_i \right) = \sum_i \mathbf{r}_i \times \mathbf{F}_i \\
&= \sum_k \sum_{i \neq k} \mathbf{r}_i \times \mathbf{F}_{ki} + \sum_i \mathbf{r}_i \times \mathbf{F}_i^{(ext)}.
\end{aligned}
\tag{8.33}
$$

The first term in the last expression does vanish for the reasons we have already explained earlier. In fact, this term has nothing to do with external forces. The second term represents the sum of the torques on the system of particles, i.e., the resultant torque. Finally, the law for the rotational dynamics can be cast on the same form which we have already observed for a single particle in Eq. (8.2),

$$\frac{d\mathbf{L}}{dt} = \sum_i \mathbf{r}_i \times \mathbf{F}_i^{(ext)} = \sum_i \vec{\tau}_i = \vec{\tau}. \tag{8.34}$$

Let us construct the analog of the impulse of the force. When an external torque $\vec{\tau} = \vec{\tau}(t)$ acts on a system of particles (or body) with angular momentum $\mathbf{L}$ (both must be taken with respect to the same point), the variation of $\mathbf{L}$ during the time period between t_1 and t_2 is given by

$$\mathbf{L}_2 - \mathbf{L}_1 = \int_{t_1}^{t_2} \vec{\tau} \, dt. \tag{8.35}$$

To better illustrate the analogy with the translational movement, consider a rigid body with axial symmetry, rotating around its axis of symmetry. The angular momentum with respect to any point on the axis is given by (8.31)

$$\mathbf{L} = I \vec{\omega}, \tag{8.36}$$

analogous to the expression of linear momentum of a body, $\mathbf{p} = m\mathbf{v}$. In addition, the formula (8.35) is similar to the Eq. (4.17).

Finally, we calculate the kinetic energy of a body with fixed axis of rotation. As we shall prove in a moment, the result is expressed by the formula analogous to the kinetic energy expression for translational movement, $K_{\text{trans}} = mv^2/2$. In accordance with the rotational analogues, we have

$$K = \frac{I\omega^2}{2}. \tag{8.37}$$

To check the validity of the expression above, remember that the relation between the angular velocity $\vec{\omega}$ and the linear velocity, given by $\mathbf{v} = \vec{\omega} \times \mathbf{r}$. This means that for a single particle

$$\mathbf{p} = m\mathbf{v} = m\left(\vec{\omega} \times \mathbf{r}\right),$$

For a particle of mass m, we can write

$$K = \frac{mv^2}{2} = \frac{m\left(\vec{\omega} \times \mathbf{r}\right)^2}{2}. \tag{8.38}$$

To work out the square of the vector $\vec{\omega} \times \mathbf{r}$, one can note that the modulus of this vector is given by

$$\left|\vec{\omega} \times \mathbf{r}\right| = \omega \cdot r \cdot \sin \alpha,$$

where α is the angle between $\vec{\omega}$ and $\mathbf{r}$. Thus, we obtain

$$\left(\vec{\omega} \times \mathbf{r}\right)^2 = \omega^2 r^2 \sin^2 \alpha.$$

Without losing generality, we can assume that the axis of rotation is the axis OZ. The quantity $r \sin \alpha$ is then exactly the distance from the rotation axis to the particle (see Fig. 8.12). Therefore, for a particle of mass m, the expression (8.38) reduces to

$$K = \frac{m\rho^2 \cdot \omega^2}{2} = \frac{I\omega^2}{2}. \tag{8.39}$$

This result can be immediately generalized to the body which consists from many points, because the kinetic energy is an additive quantity, and the same for the moment of inertia. And, by assumption, ω is the same for all matter points. Adding the contributions from all points of the body, we arrive at the relation

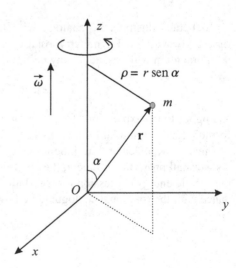

Fig. 8.12 Calculation of the kinetic energy of rotation about the axis OZ

$$K = \sum_{i=1}^{N} \frac{m_i}{2} (\vec{\omega} \times \mathbf{r}_i)^2 = \sum_{i=1}^{N} \frac{m_i}{2} \omega^2 r_i^2 \sin^2 \alpha_i = \frac{I\omega^2}{2}$$

and so we obtain again (8.39), but for the whole body.

Example 8. Calculate the kinetic energy of a homogeneous flat disc of mass M and radius R, which is rolling on a horizontal surface without slipping with the speed v (velocity of the center of mass of the disc).

Solution. We have already solved this problem in Chap. 6, for the case of a thin wheel with the mass concentrated at the circumference. The difference with the disc is only the expression for the moment of inertia, which we already calculated before,

$$I_{\text{wheel}} = MR^2, \qquad I_{\text{disc}} = \frac{1}{2}MR^2.$$

It is convenient to present the kinetic energy as a sum of the kinetic energy of the center of mass motion and the kinetic energy of internal motion, which in this case is rotation. We can write, therefore,

$$K = K_{\text{transl}} + K_{\text{rot}} = \frac{Mv^2}{2} + \frac{I_{\text{disc}} \cdot \omega^2}{2}.$$

Since $\omega = v/R$, we finally obtain

$$K = \frac{Mv^2}{2} + \frac{MR^2}{4} \cdot \frac{v^2}{R^2} = \frac{3Mv^2}{4}.$$

Note that the speed v used in relation $\omega = v/R$ is exactly equal to the velocity of the center of mass of the disc, due to the fact that the disc does not slip.

Exercises

1. The angular momentum of a system depends on the choice of the point of origin of the coordinate system. If a planet is spinning around the Sun in a circular orbit, where is the point for which the angular momentum of the planet is constant? If we calculate the angular momentum of the planet relative to another point, we see that it varies over time. Does this mean that the kinetic energy of the planet also varies in this reference frame?

Answer: The angular momentum varies for any reference point that does not belong to the axis of rotation. For any static point of reference the kinetic energy is constant.

2. A homogeneous disc is rolling without slipping from a height h, as shown in Fig. 8.13. Calculate its speed when it reaches the ground.

Answer: $v = \sqrt{\frac{4}{3}gh}$.

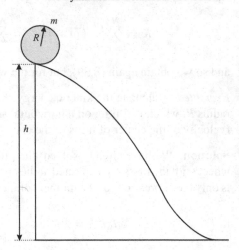

Fig. 8.13 Exercise 2

3. Solve again and generalize the Exercise 9 of Chap. 6, Sect. 6.4. Consider a wheel with mass M, radius R and moment of inertia I, rolling down on a flat surface, making the angle α with respect to the horizon. Perform the following tasks:

(a) Show that $0 \leq I \leq MR^2$. What is the distribution of mass which corresponds to the cases $I = 0$ and $I = MR^2$?
(b) Assume the wheel rolls down without slipping, and use conservation of energy to calculate linear acceleration of the center of the wheel. Consider different values of I between $I = 0$ and $I = MR^2$, and interpret the results for acceleration.
(c) Consider the case when the coefficient of friction between the wheel and the inclined plane is μ. Find the relationship (in the form of inequality) between μ and α, for which no sliding is possible. What happens with this relation when we vary I? Explain the result especially in the case $I = 0$.

Answers:

$$(b) \quad a = \frac{MR^2}{MR^2 + I} g \sin\alpha, \qquad \frac{1}{2} g \sin\alpha \leq a \leq g \sin\alpha.$$

$$(c) \quad \tan\alpha \leq \mu \left(1 + \frac{MR^2}{I}\right).$$

For I close to zero, the sliding occurs even for small angles of inclination.

4. Verify whether the formula (8.36) is valid for the situations described below, assuming rotation around a fixed axis in all cases.

   (a) A free particle with radius-vector $\mathbf{r}$ and linear momentum $\mathbf{p} = m\mathbf{v}$;
   (b) A homogeneous flat disc of mass M and radius R rotating around the axis perpendicular to its plane and passing through the center of the disc.

(c) An axially symmetric rigid body, spinning around its axis of symmetry. Consider additivity of angular momentum and moment of inertia, and use the result of item (a).

Remember that the angular momentum is an instantaneous characteristic of the movement and does not depend on the interactions between the particles.
Hint. It is desirable, in most cases, to consider the axis of rotation passing through the center of mass of the body and the origin of the coordinate system.

5. Without doing any calculations, decide whether the angular momentum (with respect to a static point) of a free particle must be constant in time. Suppose that a free particle of mass m and velocity v moves on a line, which is at a distance b from the origin. Calculate its angular momentum relative to the origin of coordinates.

6. A homogeneous disc of mass M and radius R is moving without rotation on a horizontal smooth surface, with velocity v, as shown in Fig. 8.14. Initially there is no friction. From a certain point the disc starts to move on a surface with a coefficient of friction μ. Calculate the final angular velocity of the disc and the fraction of the initial kinetic energy that is lost in the process.

Answers: $\omega = \frac{2v}{3R}$ and $\Delta K = \frac{1}{3}K_{\text{in}}$.

7. Calculate the moment of inertia of a thin homogeneous bar of mass M and length l relative to the axis passing through the center of the bar and making an angle θ with it. Determine the angular momentum vector of the bar in relation to its center of mass, assuming that the bar is rotating around the same axis with an angular velocity ω. In this case, is the equation $\mathbf{L} = I\vec{\omega}$ satisfied?

Answer: $\mathbf{L} = \frac{1}{12}Ml^2\sin\theta\,\omega\,\hat{n}$, where $\hat{n}$ is a unit vector perpendicular to the bar, which rotates together with the bar. Obviously, $\mathbf{L} \neq I\vec{\omega}$.

8. According to the text of the chapter, a rigid body which has axial symmetry and rotates around its axis of symmetry has an angular momentum $\mathbf{L} = I\vec{\omega}$. This does not mean that in order to satisfy this property, the body should possess axial symmetry. In other words, you can find examples of rigid bodies without axial symmetry that satisfy the mentioned property. What is the simplest possible configuration with

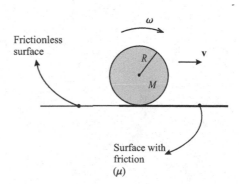

Fig. 8.14 Exercise 6

two point-like particles connected by a rigid and massless rod? What is the symmetry of the system?

9. Two blocks of equal mass m, attached to the ends of a spring with elastic constant k, form a closed system. At the initial instant, the distance between the blocks is given by $2b_0$, that corresponds to the natural length of the spring. The initial velocities of the blocks are given by $\frac{1}{2}\mathbf{v}_0$ and $-\frac{3}{2}\mathbf{v}_0$, accordingly, where the vector $\mathbf{v}_0$ has direction perpendicular to the line of the spring in its initial position. Realize the following program:

(a) Using conservation laws, determine how the velocities of the blocks in the reference frame of the center of mass (C) depend on the distance between them, $2b$;
(b) Working in the frame (C), find the maximum separation between the blocks, as well as its relationship with the minimum angular rotation speed of the system. Consider $2mv_0^2 = kb_0^2$.

Solution. The initial velocities of the blocks in the reference frame (C) are $\mathbf{v}_0$ and $-\mathbf{v}_0$. At the instant when the separation between the blocks is a maximal one, $2b$, the velocities are $\mathbf{v}$ and $-\mathbf{v}$. Then the conservation of angular momentum provides the relationship $bv = b_0 v_0$. Using the conservation of energy, we obtain

$$mv_0^2 = m\frac{v_0^2 b^2}{b_0^2} + \frac{k}{2}(b - b_0)^2.$$

Defining $b_0/b = x$, we obtain an equation for x as

$$A(1+x) = \frac{(x-1)^2}{x^2}, \qquad A = \frac{2mv_0^2}{kb_0^2} = 1.$$

The last equation can be in principle solved by Cardano's formula for algebraic equation of third-order (discovered in the sixteenth century). However, it is more practical to use numerical or graphical methods. We obtain the solution $x = 0.543689$. Then, the maximum separation value corresponds to $b = b_0/x \approx 1.84\,b_0$. The minimum angular velocity is given by the relation $\omega_m = x^2\,(v_0/b_0)$. It is interesting that the value of x does not depend on the fact that the masses are equal.

Chapter 9
Central Forces and Kepler's Laws

Abstract From the historical perspective, the derivation of three Kepler's laws in Classical Mechanics is one of the most relevant calculations which were ever done. We shall present this calculation in details and also give a simple treatment of the effect of Precession of Perihelion for a nearly circular orbit for a weakly non-Newtonian gravitational force. This problem has very special importance in General Relativity due to the Precession of Perihelion for the Mercury and some other relativistic tests.

9.1 Kepler's Laws

In what follows we consider the motion of a particle of mass m in the gravitational field produced by a point-like static body of mass M. Historically, the solution of this problem and its application to the movement of planets around a star, in the Solar System, was given by Newton and produced a great impact to the development of Classical Mechanics and Science in general. The particle of mass m is associated with a planet and the static body of mass M, to the Sun. It is clear that the same consideration can be applied to other similar systems, but we can use the terms "Sun" and "planet" to simplify the terminology.

The analysis of observational data resulted in the following astronomical laws (Kepler's Laws), discovered by Johannes Kepler (1571–1630):

1. The orbit of a planet of mass m is an ellipse with the Sun (point of mass M) lying in one of its focuses, as shown in Fig. 9.1. An important note is that, according to what has been discussed previously in Chap. 4 about the problem of two bodies, both celestial bodies (the Sun and the planet) are spinning around the center of mass of the system. The movement may be reproduced by the motion of a single body of reduced mass

$$\mu = \frac{mM}{m+M}.$$

I.L. Shapiro and G. de Berredo-Peixoto, *Lecture Notes on Newtonian Mechanics*, Undergraduate Lecture Notes in Physics, DOI 10.1007/978-1-4614-7825-6_9, © Springer Science+Business Media, LLC 2013

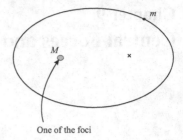

Fig. 9.1 Elliptical orbit of a
planet around the Sun

One of the foci

However, in most cases, one can use the approximation $m \ll M$, and thus the reduced
mass is very close to the mass of the lighter body (planet), i.e., $\mu \approx m$ and the
movement of the body of mass M can be neglected.

2. The area of the ellipse covered by the straight line joining the two bodies,
divided by the time interval elapsed, is a constant. In other words, the line covers
equal areas in equal times (see Fig. 9.2).

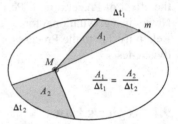

Fig. 9.2 Kepler's Second
Law: If $\Delta t_1 = \Delta t_2$, the two
areas outlined will be the
same

3. Law of the periods. If the two planets have different orbits, the ratio between
the squares of the periods of the planets, T_1 and T_2, is equal to the ratio between the
cubes of the major semi-axes of the two ellipses, a_1 and a_2. The formula looks like

$$\left(\frac{T_1}{T_2}\right)^2 = \left(\frac{a_1}{a_2}\right)^3. \tag{9.1}$$

Kepler's Laws provide an excellent fit with the astronomical observations and
represent, along with their theoretical basis (Newton's laws, as we will see shortly),
one of the largest triumphs of science. This is true regardless of the fact that these
laws are not absolutely precise. There are several reasons responsible for deviations
from Kepler's laws for the motion of planets in the Solar System. One is that the
presence of other planets and cosmic bodies makes the problem of two bodies to be
only a good approximation.

In real life, the Sun interacts with all planets of the Solar system and the situation is much more complicated. The present-day observational possibilities enable
astronomers to describe the motion of the whole Solar system, but this requires the
use of numerical methods. It is good to know that the dynamics of the gravitating
systems with more than two bodies has no periodicity and moreover leads to the
phenomenon called chaos. In simple terms, this means that even a small uncertainty

in the initial data will grow with time. As a result it is absolutely impossible, for instance, to say whether in the given place of Earth, after a million years (measured by atomic clocks) there will be day or night, or even winter or summer. Analytical periodic solution is possible only for the two-body problem which we will consider in what follows.

Furthermore, the Sun itself rotates around its axis, and its shape is not exactly spherical. As a result, the gravitational field produced by the Sun is not exactly the same field generated by a point-like or spherically symmetric mass. As we will see throughout this chapter, Kepler's Laws correspond to the point-like mass approximation. Finally, there are also relativistic effects, so that the correct gravitational potential is slightly different from the Newtonian potential. We will return to discuss the last type of deviation from Kepler's Laws later on. For now, we will concentrate our attention on the derivation of Kepler's Laws.

Our purpose is to get the Laws of Kepler from Newton's laws and their consequences, such as the conservation laws. As exercises, we suggest other applications of Newton's laws for the motion in a Newtonian gravitational potential, namely the motion of a cosmic body with the speed higher than the escape velocity, as a consequence, such a particle has an open orbit. Furthermore, we will consider the deviations from Kepler's Laws for closed movements, including the calculation of the precession of the perihelion of Mercury, a phenomenon that represents one of the most important tests of General Relativity.

The first observation is that the second law of Kepler is equivalent to the conservation of angular momentum. Let us explain this in detail. As can be seen by the triangle shown in Fig. 9.3, the sectorial velocity (also called areal velocity) is given by

$$A' = \frac{dA}{dt} = \frac{1}{2} |\mathbf{r}| \cdot \left| \frac{d\mathbf{r}}{dt} \right| \sin \alpha = \frac{1}{2} |\mathbf{r} \times \mathbf{v}| = \frac{1}{2m} |\mathbf{L}| . \tag{9.2}$$

According to Eq. (9.2), Kepler's Second Law simply means that the angular momentum is a constant of motion. Considering the movement of a point-like body of mass m in a central field, or more specifically, in the gravitational field generated by a point-like body of mass M, we can write the force exerted on the test particle in the form[1]

$$\mathbf{F} = -\frac{GMm}{r^3} \mathbf{r}. \tag{9.3}$$

This force produces zero torque:

$$\vec{\tau} = \mathbf{r} \times \mathbf{F} = 0.$$

---

[1] Let us remember that a central force field is characterized by the spherical symmetry about the origin (force acts in the radial direction and its modulus is independent of angular coordinates). The gravitational field generated by a spherically symmetric distribution is a central field.

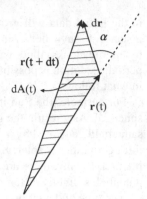

Fig. 9.3 Calculation of the area swept in infinitesimal time

Then, according to the law of rotational dynamics, $\dot{\mathbf{L}} = \vec{\tau} = 0$ and, consequently, $\mathbf{L} = const$. According to the Eq. (9.2), this implies $A' = constant$, i.e., we have demonstrated the second Kepler's Law.

By the considerations presented above, it is obvious that the second Kepler's Law applies to the movement in any kind of central field, $\mathbf{F} = F(r)\hat{\mathbf{r}}$. On the other hand, other Kepler's Laws require a specific form of potential, as we shall see in what follows.

Before discussing the first and third laws, we consider the simplest particular case of a circular motion of a particle of mass m. In this case we have $\mathbf{v} \perp \mathbf{r}$ in all points of the trajectory, as well as $v = \omega r$, where $\omega = const$.

Newton's Second Law provides

$$\frac{mv^2}{r} = m\omega^2 r = \frac{GMm}{r^2},\tag{9.4}$$

such that $\omega^2 = GM/r^3$. Recalling that the period of rotation is expressed in terms of angular frequency by the ratio $T = 2\pi/\omega$, we can write

$$\frac{T^2}{(2\pi)^2} = \frac{r^3}{GM}.$$

This is the Third Law for the periods of circular motion. Such a motion requires a fine adjustment of the magnitude and direction of the initial velocity $\mathbf{v}_0$. Since the initial position is defined by $\mathbf{r}_0$, the two vectors, $\mathbf{v}_0$ and $\mathbf{r}_0$, must be mutually perpendicular and the value of v_0 adjusted by the relationship (9.4). It is clear that for the general case, where all these fine-tunings do not take place, we still have to prove the Kepler's Third Law.

To analyze the general case, we have to choose the constant parameters characterizing the motion of the point-like body of mass m. We know that the mechanical energy of the particle E and its angular momentum $\mathbf{L}$ are constants of motion, because the force of our interest is central and, in particular, given by (9.3). So, let us consider a movement of a body with energy E and angular momentum $\mathbf{L}$.

What are the physical consequences of these conservation laws? $\mathbf{L} = constant$ has the following consequences:

(a) We know $\mathbf{L} = \mathbf{r} \times \mathbf{p}$, hence $\mathbf{L} \perp \mathbf{r}$ and $\mathbf{L} \perp \mathbf{p}$ for all points. This means that the movement occurs in the same fixed plane, perpendicular to the vector $\mathbf{L}$;

(b) The modulus of the angular momentum, $L = |\mathbf{r} \times \mathbf{p}| = rmv \sin \alpha$, is constant.

Already the item (a) leads to a great simplification of dynamical problem of our interest. In fact, the problem which was three-dimensional, now became a two-dimensional one. For a description of the motion on a plane, we can use polar coordinates with the origin in the center of force, where we set the particle of mass M. Other consequences of $\mathbf{L} = constant$ will be discussed later on.

In order to use energy conservation, we write the formula for kinetic energy of the particle in polar coordinates. As we have learned in Chap. 2, this can be done as follows:

$$K = \frac{mv^2}{2} = \frac{m\left(\dot{r}^2 + r^2\dot{\varphi}^2\right)}{2}. \tag{9.5}$$

Looking at Fig. 9.3, one can note that $d\mathbf{r}$ has the same direction as the velocity vector. Thus, $v \sin \alpha = r\dot{\varphi}$ and hence $L = mr^2\dot{\varphi}$. As far as $\mathbf{L}$ is an integral of motion and its modulus depends only on the initial data, we arrive at the important relation between angular velocity and momentum,

$$\dot{\varphi} = \frac{L}{mr^2}, \tag{9.6}$$

where $L = const$. The last equation relates the angular velocity of the planet with the distance between the body and the center, i.e., the position of the body of mass M.

Now we can use the relationship (9.6) in the expression for the kinetic energy (9.5), eliminating the angular velocity,

$$K = \frac{m\dot{r}^2}{2} + \frac{mr^2}{2} \cdot \frac{L^2}{m^2r^4} = \frac{m\dot{r}^2}{2} + \frac{L^2}{2mr^2}. \tag{9.7}$$

An important feature of the last formula is that the new expression for kinetic energy depends only on the variable $r(t)$ and its time derivative, $\dot{r}(t)$, regardless of the angular coordinate φ. Note, however, that the dynamics of angular motion exists, being described by the relation (9.6). The knowledge of the function $r(t)$ together with the relation (9.6) enables one to determine $\varphi(t)$.

As we have already stated, the kinetic energy (9.7) depends only on r. Moreover, we know that the potential energy of interaction between two bodies also depends only on r. For the Kepler problem, for gravitational interaction, we have

$$U(\mathbf{r}) = U(r) = -\frac{\alpha}{r}, \qquad \text{where} \qquad \alpha = GMm. \tag{9.8}$$

Then the law of energy conservation can be written as

$$E = \frac{m\dot{r}^2}{2} - \frac{\alpha}{r} + \frac{L^2}{2mr^2} = \frac{m\dot{r}^2}{2} + U_{\text{eff}}(r), \tag{9.9}$$

where $E = constant$ and

$$U_{\text{eff}}(r) = -\frac{\alpha}{r} + \frac{L^2}{2mr^2} \qquad (9.10)$$

is oftenly called *effective potential energy*. Actually, this is the correct formula for potential energy if we consider only the radial movement. Note that the purely radial movement can only be observed in the reference frame rotating with angular velocity $\dot{\varphi}$ relative to the inertial frame (laboratory). But in the rotating reference frame there are the inertial forces, and in particular, there is the centrifugal force. The term $\frac{L^2}{2mr^2}$ can be interpreted, in this case, as the potential energy associated to the centrifugal force. The replacement of the potential energy (9.8) by the effective potential energy (9.9) can be seen as a procedure equivalent to a transformation from the inertial frame to the accelerated frame.

As far as in (9.9) there is only one variable, $r(t)$, we have effectively reduced the two-dimensional dynamical problem to a one-dimensional description. Now we can use the method developed in the previous chapters to study the motion of a point-like body in a fixed potential of external central force. It proves useful to classify motions according to the value of the total mechanical energy of the system. The graph of $U_{\text{eff}}(r)$ is shown in Fig. 9.4. Depending on the value of the energy, the system can perform either bounded (limited) or unbounded movements.

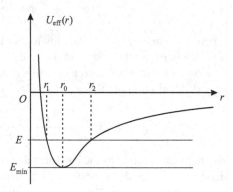

Fig. 9.4 Graph of the effective potential

If the system has the minimum possible energy, E_{min}, the motion will be characterized by $r = r_0$, i.e., this is a circular motion. Let us calculate r_0 and E_{min}. Taking the derivative of the effective potential, we obtain

$$U'_{\text{eff}}(r) = -\frac{L^2}{mr^3} + \frac{\alpha}{r^2}.$$

Then,

$$U'_{\text{eff}}(r_0) = 0 \qquad \Longrightarrow \qquad r_0 = \frac{L^2}{\alpha m}. \qquad (9.11)$$

We also obtain, by replacing this value back into potential,

$$E_{\min} = -\frac{\alpha^2 m}{2L^2}.$$

(9.12)

It is easy to verify that these are the values for energy and radius corresponding to a circular motion, i.e., the solution of Eq. (9.4). We leave this consideration as an exercise.

To interpret and understand the motions in the effective potential, we consider different values of energy E.

1) $E < E_{\min}$. In this case, the motion is impossible, because it implies a negative radial kinetic energy, i.e., $\dot{r}^2 < 0$.

2) $E = E_{\min}$. In this case, the planet (test particle) cannot have non-zero radial speed, because the only allowed value for the radial coordinate is $r = r_0 = const$, according to (9.11). The movement is circular (two-dimensional) with a constant angular velocity given by (9.6).

3) $E_{\min} < E < 0$. For these values of energy, the planet performs a movement which is radially limited, but more complicated than the circular motion. In particular, there are two turning points, r_1 and r_2, and the particle moves in this range ($r_1 < r < r_2$). As we show below, the motion (always two-dimensional) occurs in a closed path in the form of an ellipse, according to Kepler's Laws.

4) $E \geq 0$. In this case the planet escapes into infinity, i.e., its initial speed is greater than the escape velocity. For $E = 0$, the trajectory is a parabola and for $E > 0$, it is a hyperbola. We leave the analysis of these cases as an exercise.

We consider in details only the case (3), where $E_{\min} < E < 0$. We have the following equations for the description of dynamics: the Eq. (9.6) and the equation for the radial movement, which follows directly from Eq. (9.9),

$$\dot{r}^2 = \frac{2}{m}\left[E - U_{\text{eff}}(r)\right].$$

Here E and L are constants. We can cast the two equations into a more useful form,

$$\frac{dr}{dt} = \pm\sqrt{\frac{2}{m}}\sqrt{E - U_{\text{eff}}(r)} \quad \text{and} \quad \frac{d\varphi}{dt} = \frac{L}{mr^2}.$$

(9.13)

If we are interested only in the trajectory of motion, it will be convenient to express the time derivatives in the first of equations (9.13) in terms of derivative with respect to the angle φ. In this way we obtain

$$\frac{dr}{d\varphi} = \frac{dr}{dt}\cdot\frac{dt}{d\varphi} = \pm\frac{mr^2}{L}\cdot\sqrt{\frac{2}{m}}\left[E + \frac{\alpha}{r} - \frac{L^2}{2mr^2}\right]^{1/2},$$

i.e.,

$$-\frac{dr}{r^2} \cdot \left(E + \frac{\alpha}{r} - \frac{L^2}{2mr^2}\right)^{-1/2} = \pm \frac{\sqrt{2m}\,d\varphi}{L}.$$

Introducing the useful notation $u = 1/r$, we rewrite the last equation as

$$\frac{du}{\sqrt{E + \alpha u - \beta u^2}} = \pm \frac{\sqrt{2m}\,d\varphi}{L}, \quad \text{where} \quad \beta = \frac{L^2}{2m}. \tag{9.14}$$

In this equation the variables are already separated and one needs just to integrate the two sides. In order to integrate the left hand side, we can write the argument of the square root as

$$E + \alpha u - \beta u^2 = -\beta\left(u - \frac{\alpha}{2\beta}\right)^2 + \frac{\alpha^2}{4\beta} + E$$

and note that $\frac{\alpha^2}{2\beta} + E \geq 0$ for $E \geq E_{\min}$. The Eq. (9.14) can be, after that, integrated in the form

$$\int \frac{d\left(u - \frac{\alpha}{2\beta}\right)}{\sqrt{-\beta\left(u - \frac{\alpha}{2\beta}\right)^2 + \frac{\alpha^2}{4\beta} + E}} = \pm \frac{\sqrt{2m}\,(\varphi - \varphi_0)}{L}, \tag{9.15}$$

where φ_0 is the polar angle corresponding to the turning point r_1. To simplify the notations, we can choose the orientation of the axes OX and OY such that $\varphi_0 = 0$. The integration is simple and the result is given by (see more details in Exercise 3)

$$\cos\varphi = \frac{\frac{L}{r} - \frac{m\alpha}{L}}{\sqrt{2mE + \frac{m^2\alpha^2}{L^2}}}. \tag{9.16}$$

The last formula is the solution we were looking for. The next step is to interpret it geometrically. The first observation is that the signs $\pm$ in Eq. (9.15) correspond to the choice of direction of rotation and thus have no relevance for defining the trajectory. As we shall see, Eq. (9.16) is the equation of an ellipse. Let us introduce the standard geometric notations

$$p = \frac{L^2}{m\alpha}, \tag{9.17}$$

called parameter of the ellipse, and

$$e = \sqrt{1 + \frac{2EL^2}{m\alpha^2}} < 1, \tag{9.18}$$

called the eccentricity of the ellipse. Using these notations, relation (9.16) assumes a more compact form

$$\frac{p}{r} = 1 + e\cos\varphi. \tag{9.19}$$

Equation (9.19) is the canonical equation of an ellipse in polar coordinates. But it is also interesting to present this solution in Cartesian coordinates. We can rewrite the last equation using $x = r\cos\varphi$ and $y = r\sin\varphi$ in the form

$$x^2 + y^2 = p^2 - 2pex + e^2x^2.$$

Finally, after a small algebra we arrive at the equation

$$\frac{(1-e^2)^2}{p^2}(x-x_0)^2 + \frac{1-e^2}{p^2}y^2 = 1,$$

where

$$x_0 = \frac{ep}{e^2 - 1}.$$

We have then the following equation of an ellipse in Cartesian coordinates (see Fig. 9.5):

$$\frac{(x-x_0)^2}{a^2} + \frac{y^2}{b^2} = 1,$$

with

$$a = \frac{p}{1-e^2} = \frac{L^2}{m\alpha} \cdot \frac{m\alpha^2}{2|E|L^2} = \frac{\alpha}{2|E|},$$

$$b = \frac{p}{\sqrt{1-e^2}} = \frac{L^2}{m\alpha} \cdot \frac{\sqrt{m\alpha}}{\sqrt{2|E|L^2}} = \frac{L}{\sqrt{2m|E|}}. \tag{9.20}$$

The position of the center of the ellipse corresponds to $x_0 = ea$, while the Sun (static center of mass M) is in one of the focus of the ellipse. Therefore, we conclude that the first law of Kepler is demonstrated.

Finally, to demonstrate the law of periods, remember that the area covered per unit of time, $dA/dt = A'$, is constant and $L = 2mA'$. As far as the area of the whole ellipse is equal to πab, one can obtain the period of the planet's orbit around the Sun simply by dividing the area by the corresponding rate A',

$$T = \frac{\pi ab}{A'} = \frac{2\pi p^2 m}{(1-e^2)^{3/2}|L|} = \frac{2\pi\alpha|L|m}{2|E||L|\sqrt{2m|E|}}$$

$$= \alpha\pi\sqrt{\frac{m}{2|E|^3}} = 2\pi\sqrt{\frac{m}{\alpha}} \cdot a^{3/2}. \tag{9.21}$$

This is indeed the Kepler's Third Law (law of periods). One can note that the period, T, is independent of the magnitude of the angular momentum L and is related only to the energy E. More information about the solution of Kepler's problem can be found, e.g., in the book [21].

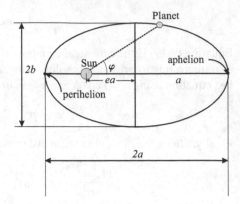

Fig. 9.5 Elliptical orbit of a planet around the Sun for $E_{min} < E \leq 0$. The Sun is at one of the focus of the ellipse

Exercises

1. Show that, at least for a circular motion, the term

$$\Delta U = \frac{L^2}{2mr^2},$$

of Eq. (9.9), is exactly the potential energy associated with the centrifugal force, $\mathbf{F}_c$.

Hint. Consider the motion in a rotating reference frame with angular velocity given by (9.6) and show that $\mathbf{F}_c = -\operatorname{grad}\Delta U$.

2. Verify that the minimum energy value (9.12) and the value of the radial coordinate (9.11) correspond to the circular motion, i.e., the solution of the Eq. (9.4). Check that E_{min} corresponds exactly to the sum $K + U$ in the case of circular movement.

3. Develop the calculation of the integral (9.16).

Solution: We write the left part of Eq. (9.14) as

$$\frac{1}{\sqrt{\frac{\alpha^2}{4\beta} + E}} \cdot \frac{du}{\sqrt{1 - \gamma(u - \frac{\alpha}{2\beta})^2}}, \qquad \gamma = \frac{4\beta^2}{\alpha^2 + 4\beta E}.$$

As already considered, $E > E_{min} = -\frac{\alpha^2}{4\beta}$, then we always have $\gamma > 0$. Recalling that $\beta = L/(2m)$, we come to

$$\pm d\varphi = d \arccos\left\{\frac{2\beta}{\sqrt{\alpha^2 + 4\beta E}} \cdot \left(u - \frac{\alpha}{2\beta}\right)\right\},$$

i.e.,

$$\pm \varphi = \arccos\left\{\frac{2\beta}{\sqrt{\alpha^2+4\beta E}}\left(u-\frac{\alpha}{2\beta}\right)\right\} = \arccos\left\{\frac{L^2/m}{\sqrt{\alpha^2+\frac{2EL^2}{2m}}}\left(\frac{1}{r}-\frac{m\alpha}{L^2}\right)\right\},$$

which is equivalent to the relationship (9.16).

4. For the elliptical motion, calculate the minimum, r_{min}, and maximum, r_{max}, distances from a planet to the Sun at $r = 0$. The points correspond to the perihelion and the aphelion of the elliptical orbit. Check that these values match the turning points in the radial motion.

Answers: $r_{min} = a(1-e)$ and $r_{max} = a(1+e)$, where the constant a is defined by the Eq. (9.20) and e is the eccentricity defined in Eq. (9.18).

5. Consider the motion for $E \geq 0$. Show that the trajectory is a hyperbola in the case $E > 0$ and parabola in the case $E = 0$. For $E > 0$ calculate the deviation angle of the body, θ, as a function of the impact parameter of this scattering problem. The impact parameter h is defined as follows: initially the body of mass m is moving along a straight line, and h is the distance between this line and the center of mass M (see Fig. 9.6).

Hint and Answers: As a first step, one has to check that formula (9.19) is also valid for $E \geq 0$, with eccentricity $e = 1$ in the case $E = 0$ and $e > 1$ for $E > 0$. The calculations for the case $e = 1$ are quite simple and for $E > 0$ the equation of the trajectory is given by

$$\frac{(x-x_0)^2}{a^2} - \frac{y^2}{b^2} = 1, \qquad x_0 = \frac{ep}{e^2-1} > 0, \qquad (9.22)$$

with

$$a = \frac{p}{e^2-1}, \qquad b = \frac{p}{\sqrt{e^2-1}}. \qquad (9.23)$$

which is a natural modification of Eq. (9.20). To calculate the angle of deviation, we have to write the energy and the angular momentum in terms of the parameter of impact and the initial velocity of the particle v_0. The angle between the asymptotic line $ay = \pm b(x-x_0)$ and the axis OX is given by $\varphi = \arctan\sqrt{e^2-1}$. Furthermore, for the parameter of impact, we have $h = x_0 \sin \varphi$. Using these results, we obtain

$$\theta = \pi - 2\varphi = \pi - 2\arctan\left(\frac{2Eh}{\alpha}\right). \qquad (9.24)$$

6. Solve the previous problem for the case of a repulsive force, considering $\alpha < 0$.

Answer: The movement is possible only for $E \geq 0$. The equation of the trajectory, in polar coordinates, is given by the equation

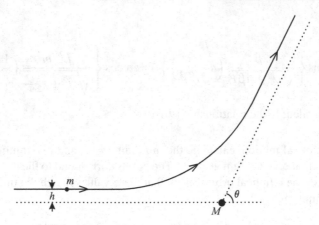

Fig. 9.6 Scattering problem – Exercise 5

$$\frac{p}{r} = -1 + e\cos\varphi.$$

The main difference is that in this case $x_0 > 0$ in the equation of the trajectory, (9.22). The relation between the impact parameter h and the deviation angle has a similar form to Eq. (9.24),

$$\theta = \pi - 2\varphi = \pi - 2\arctan\left(\frac{2Eh}{|\alpha|}\right). \tag{9.25}$$

7. Consider the motion of a planet in the case of an ellipse with a very small eccentricity, i.e., for a trajectory very close to a circle. Separating the radial movement in the same manner as done in the text of this section, consider the small oscillations around the equilibrium position, $r = r_0$. Show that the period of the harmonic oscillations in radial direction is equal to the period of rotation. That means that the planet's orbit remains closed (in the first approximation) in the case of a small eccentricity. In the next section, we use this method to calculate the precession of the perihelion of Mercury due to the weak relativistic effects.

Solution: As we already know, the effective potential energy (9.10) has its point of minimum at $r_0 = \frac{L^2}{m\alpha}$. Consider, for the movement in the radial direction, $r(t) = r_0 + q(t)$, where $q(t)$ is a small perturbation. In this case, the energy of the harmonic oscillator can be obtained by the linearization of Eq. (9.9),

$$E_{lin} = \frac{m\dot{q}^2}{2} + \frac{kq^2}{2}, \qquad \text{where} \qquad k = \frac{m^3\,\alpha^4}{L^6}, \tag{9.26}$$

so that the frequency of radial oscillation is given by $\omega = \sqrt{k/m} = m\alpha^2/L^3$.

On the other hand, the frequency of the circular movement is defined as $\omega = v/r$, where $mr_0 v = L$. We can see that the two frequencies coincide, demonstrating that the trajectory is indeed closed, at least in the linear approximation. It is interesting to know that in Mathematical Physics the same method is used in the search of potentials $U(r)$ that guarantee closed orbits. For now, beyond the Kepler's problem, it is known only one more example of such potential.

8. The Laplace-Runge-Lenz vector is defined as

$$\mathbf{A} = \mathbf{p} \times \mathbf{L} - m\alpha\hat{\mathbf{r}}, \tag{9.27}$$

where the Newton force acting on the particle (planet) is

$$\mathbf{F} = -\frac{\alpha}{r^2}\hat{\mathbf{r}}.$$

Show that (a) Vector $\mathbf{A}$ is a constant of motion, $d\mathbf{A}/dt = 0$. (b) There is a relation between the modulus of this vector and the eccentricity of the orbit, $|\mathbf{A}| = m\alpha e$.

9.2 Precession of Perihelion for a Nearly Circular Orbit

In this section we consider the dynamics of a particle in a central field with potential $U(r)$, assuming a small deviation from the law of gravitation of Newton, namely

$$U(r) = U_N(r) + \Delta U(r),$$
$$\text{where} \quad U_N(r) = -\frac{\alpha}{r} \quad \alpha = GMm. \tag{9.28}$$

We are interested in situations when the correction $\Delta U(r)$ is small. The main effect of the correction to the Newtonian potential should be as follows. As we have seen in the previous section of this chapter, in the case of Newtonian potential the motion of a particle (e.g., a planet) satisfies the First Law of Kepler. This means, in particular, that the trajectory of the particle is a closed curve. In other words, after a certain period of time the particle returns to its initial position. In accordance to the laws of Kepler, the trajectory is an ellipse. In the case of the planets of the Solar System, in a first approximation, Kepler's laws are satisfied, but if we increase the accuracy of astronomical measurement, small deviations may be observed. The main manifestation of these deviations is the fact that the orbits of the planets are no longer closed, i.e., they are not exactly elliptic. The orbits of the planets are very close to circular paths and so it is very difficult to observe the mentioned deviations. The planet which has the largest eccentricity is Mercury ($e = 0.2056$) while, for example, the eccentricity of Earth's orbit is given by $e = 0.0167$ and of Venus is $e = 0.0068$. For this reason, the orbit of Mercury is the preferred object to explore deviations from Kepler's Law. As it was already mentioned above, the difference $\Delta U(r)$ between the real potential and the Newton potential is small. Therefore, we can describe the motion of a planet of mass m as an ellipse that is slowly rotating. This process is called precession of the perihelion of the orbit of this planet. In the case of Mercury, the speed of such a precession is given by about $575''$ for a century.

From this amount, $532''$ are explained by the disturbances generated by the effects of other planets plus other effects (see the beginning of the chapter), and the rest $43''$ that are left may indicate a change in the potential according to

$$\Delta U(r) = -\frac{\beta}{r^3}, \qquad \beta = \frac{GML^2}{mc^2}. \tag{9.29}$$

Here L is the magnitude of the angular momentum of the particle of mass m. In the expression (9.29), c is the speed of light that must be considered finite so that we can identify β with small relativistic corrections. To display the formula (9.29) in a symmetric manner, we remember that $L = hm$, where h does not depend on the mass of the test particle m. Using this notation, we obtain $\beta = GmMh^2/c^2$.

The origin of the modification (9.29) is the theory of General Relativity of Albert Einstein. This theory is a relativistic theory of gravitation, which almost coincides with the Newton theory of gravitation in the limit of small velocities and weak gravitational fields. As a relativistic theory of gravitation, General Relativity is definitely outside the scope of our consideration. At the same time, concerning the motion of the planets of the Solar System, General Relativity allows us to establish small corrections to the Newton's Law. After that the theory with these corrections can be fully considered in the context of Classical Mechanics.

The treatment of the general case can be found in the books of General Relativity, here we intend to simplify the task by considering only the case of a small eccentricity, when the orbit of the planet is close to a circle. In this case, the derivation of the precession of the perihelion of Mercury becomes very simple and can be reduced to the analysis of a harmonic oscillator.

The equations describing the dynamics of the system are given by

$$mr^2\frac{d\varphi}{d\tau} = L,$$
$$\frac{m}{2}\left(\frac{dr}{d\tau}\right)^2 + \frac{mr^2}{2}\left(\frac{d\varphi}{d\tau}\right)^2 + U(r) = E. \tag{9.30}$$

The parameter τ is called *proper time*, a typical concept of relativistic theories. However, for a slow movement, such as the one of a planet, τ is very close to the usual physical time t. As we shall see shortly, the difference between τ and t produces no influence on the trajectory and hence is irrelevant for our consideration.

It is easy to identify $E = constant$ with the mechanical energy of the planet and the first equation in (9.30) is exactly the law of conservation of the angular momentum L. Using this law in the second equation, we obtain

$$mr^2\frac{d\varphi}{d\tau} = L, \qquad \left(\frac{dr}{d\tau}\right)^2 + \frac{L^2}{m^2r^2} + \frac{2U(r)}{m} = \frac{2E}{m}. \tag{9.31}$$

Now, we can eliminate the proper time, just as we eliminated the common time in the previous section. Thus, the one-dimensional equation for the trajectory of the planet has the following form:

$$\frac{L^2}{m^2 r^4}\left(\frac{dr}{d\varphi}\right)^2 = \frac{2}{m}\left[E - U_{\text{eff}}(r)\right],$$

$$U_{\text{eff}}(r) = U(r) + \frac{L^2}{2mr^2}, \tag{9.32}$$

where the function $U(r)$ is defined in (9.29).

It proves useful to rewrite equation (9.32) in a more usual form, similar to the one-dimensional dynamics of a particle. For this end we introduce (just as when solving the Kepler problem) the new variable $u = 1/r$ and write

$$\left(\frac{du}{d\varphi}\right)^2 = \frac{2}{\mu}\left[E - V(u)\right],$$

$$\text{with} \quad \mu = \frac{L^2}{m}; \quad V(u) = \frac{\mu u^2}{2} - \alpha u - \beta u^3. \tag{9.33}$$

We note that, mathematically, the last relation is a formula for the movement of a particle on the line. However, it is obvious that the physical sense of all quantities here is quite different.

For example, $\mu = L^2/m$ plays the role of a mass, φ – a role of time, u is the new coordinate on the line and $V(u)$ is the "potential energy" of the "particle". Of course, we remember that, from a physical viewpoint, Eq. (9.33) describes the dependence between the radial u and angular φ coordinates, i.e., this is a differential equation for the trajectory of the planet. However, mathematically the formula (9.33) describes a particle on the line. So, regardless on the difference in physical interpretation, we can use the approach developed in Chap. 7 to explore small fluctuations and finally analyze the trajectory.

The last term in $V(u)$ is a relativistic correction, and the rest does not represent anything new compared to the problem of Kepler. Therefore, the strategy for solving the problem consists in the following steps:

1) Find circular orbit that corresponds to the minimum of the potential energy $V(u)$.

2) Consider small radial oscillations near this orbit and find the frequency of these oscillations ω.

3) Calculate the period of these oscillations, $\Phi = 2\pi/\omega$. Let us remember that our "time" here is, in fact, the angle of rotation. So if the period is given by $\Phi = 2\pi$, the orbit is closed and there will be no precession. Correspondingly, a small angular deviation from 2π in the period indicates a slow precession of the elliptical orbit.

Let us first apply this scheme to the Newtonian case, for the purpose of training. In this situation, instead of (9.33), we have the reduced potential

$$V_N(u) = \frac{\mu u^2}{2} - \alpha u. \tag{9.34}$$

For the radius of the circular orbit u_0, we find the equation

$$\frac{dV_N(u)}{du}\bigg|_{u=u_0^N} = \mu u_0^N - \alpha = 0,$$

$$\text{then,} \qquad u_0^N = \frac{\alpha}{\mu}. \tag{9.35}$$

Now we can calculate the square of the frequency in the Newtonian case, let's call it ω_N. For this end we need to divide the elastic constant of the oscillator by its mass,

$$\omega_N^2 = \frac{1}{\mu} \cdot \frac{d^2 V_N(u)}{du^2}\bigg|_{u=u_0^N} = 1, \tag{9.36}$$

then $\omega_N = 1$ and $\Phi = 2\pi$. This means that the period of oscillation in the radial direction coincides exactly with the period of angular momentum. As one should definitely expect, there is no precession in the Newtonian case.

In the general case, we arrive at the following equation for the radius of the circular orbit and the point of minima of the potential u_0:

$$\frac{dV(u)}{du}\bigg|_{u=u_0} = -\alpha + \mu u_0 - 3\beta u_0^2 = 0. \tag{9.37}$$

Let us remember that the relativistic correction is supposed to be small. For this reason we can treat the term with β as a small perturbation and thus write the solution in the form $u_0 = u_0^N + \Delta u_0$, where Δu_0 is a small correction. In the first order in β we obtain

$$\Delta u_0 = 3\beta (u_0^N)^2 = \frac{3\alpha^2 \beta}{\mu^2}.$$

At this point we can replace the value of u_0 into the formula for the frequency,

$$\omega^2 = \frac{1}{\mu} \cdot \frac{d^2 V(u)}{du^2}\bigg|_{u=u_0} = 1 - \frac{6\alpha\beta}{\mu^2}.$$

In this formula one can observe a small correction to the Newtonian frequency, given by Eq. (9.36). The period of oscillation in the radial direction is given by

$$\Phi = \frac{2\pi}{\omega} = 2\pi \left(1 + \frac{3\alpha\beta}{\mu^2}\right) = 2\pi + \frac{6\pi G^2 m^2 M^2}{c^2 L^2}, \tag{9.38}$$

where we have used the definition (9.29) and the expansion

$$\frac{1}{\sqrt{1+x}} \approx 1 - \frac{x}{2} \qquad \text{for small} \qquad x.$$

The result (9.38) indicates that, for each period of radial oscillation, the planet rotates at an angle slightly greater than 2π. The more exact formula of General

Relativity is very close to (9.38). The only difference is an extra factor, $(1 - e)^{-1}$, related to the eccentricity of the orbit e. The result (9.38) is in the excellent accordance with the deviation of $43''$ mentioned in the beginning of the section. This result represents one of the major tests of General Relativity. Here we could reproduce a basic part of this result in an very economic way compared to the complete discussion which can be found in the textbooks on General Relativity, (e.g., [23]).

Exercises

1. Use an alternative approach to analyze the Eqs. (9.30), considering the time dependence of r.

Hints. Technically, this method is more complicated and hence the problem can be seen as a calculational challenge for a student. One has to realize the following program:
Using the second equation in (9.31), find the point of equilibrium $r_0 = r_{0N} + \delta r_0$. It can be seen that $\delta r_0 = -3m^2 MG$. Then find the frequency of small oscillations of the radial coordinate $\omega_r = \omega_{rN} + \delta \omega_r$ near the position of the equilibrium. The values r_{0N} and ω_{rN} correspond to the non-disturbed potential, $U_N = -GMm/r$. Now find the period of the radial oscillations and multiply it by the angular velocity of the relations (9.31). Remember that the angular velocity can also suffer a change because of $\delta \omega_r$. Check that the result for the precession of the perihelion is the same found in (9.38).

2. Repeat the analysis performed in this section for another form of correction to the background (e.g., Newtonian) potential. Consider $U(r) = U_0(r) + U_1(r)$, where $U_1(r)$ is a small correction to the main part of the potential, $U_0(r)$. We define r_0 as a stationary point for the effective potential without the term $U_1(r)$ and $r_1 = r_0 + \delta r$ as a stationary point for the effective potential with $U_1(r)$. The goal of the investigation is to find the correction to the frequency, caused by the presence of $U_1(r)$.

Hints and Answers:

$$\text{We define} \quad V_0(u) = U_0(r) + \frac{\mu u^2}{2}, \quad V_1(u) = U_1(r),$$

We have, for the non-disturbed potential, $V_0'(u_0) = 0$, $V_0''(u_0) = k_0$, $\omega_0^2 = k_0/\mu$. In the disturbed case, we obtain

$$k = k_0 + \delta k, \quad \delta k = V_1''(u_0) - \frac{V_0'''(u_0) \cdot V_1'(u_0)}{V_0''(u_0)}$$

and, finally,

$$\omega = \omega_0 \left(1 + \frac{\delta k}{2 k_0}\right).$$

Chapter 10
Basic Notions of Hydrodynamics

Abstract Our goal is to present a very brief introduction to the basics of Hydrodynamics. Namely, we will introduce the notions of ideal fluid, Pascal's law and the continuity equation. Furthermore, we shall obtain Euler's equation as an analog of Newton's second law for the ideal fluid and consider some consequences, such as Bernoulli's equation, which can be seen as the continuous version of energy conservation. Finally, we will discuss the equation for sound waves, representing a propagating perturbation of density and pressure.

10.1 Introduction

In previous chapters we have considered the dynamics of mechanical systems with a finite number of degrees of freedom (independent coordinates). Even when we dealt with a body of finite size, it was always sufficient to take a few coordinates. For example, in case of a rigid body one can use three Cartesian coordinates for the center of mass of the body and three angles to describe its orientation. This is possible because the coordinates of the rigid body are constrained, namely, the distances between each two points of the body are fixed and treated as constants. From a physical viewpoint this means we assume that the deformations of the body are so small that we can neglect them. In other words, we assume that a "rigid" body is infinitely rigid. There are, of course, other physically interesting cases for which such constraints do not hold. Typically, this is so for real solid bodies, fluids or fields. In all these cases one has to deal with the dynamics of the systems of different sorts, which require a continuous description, where the coordinates of each point of the body or the media are not constrained. In particular, fields and fluids require an infinite number of degrees of freedom for their description.

In the present chapter we are going to consider only the case of fluids, and leave such interesting subjects as the theory of elasticity in solids and theory of fields aside. In part, it is because a reasonable exposition of the theory of elasticity requires a bit more complicated mathematical instrument (tensor calculus) than the one we

I.L. Shapiro and G. de Berredo-Peixoto, *Lecture Notes on Newtonian Mechanics*, Undergraduate Lecture Notes in Physics, DOI 10.1007/978-1-4614-7825-6_10, © Springer Science+Business Media, LLC 2013

assume here. On the other hand, the dynamics of fields is much better presented in the framework of Lagrangian mechanics and also it is much more naturally introduced in a course on electromagnetism, and therefore it goes beyond the scope of the present book.

The theory of fluids (hydro- and aerodynamics) is very rich from both the theoretical side and in view of applications which extend from Mechanical Engineering to Nuclear Physics, Plasma Physics and Cosmology. In general, Hydrodynamics is a very important part of modern Physics and it seems important to introduce some basic notions even in a very introductory course. Here we intend to restrict our attention to the most simple aspects of the theory. In the present chapter we consider a particular example, namely the dynamics of the ideal non-relativistic fluid. The term ideal means an approximation which disregards dissipation of energy into heat, such that one can ignore the thermal properties of the fluid. In the next sections we shall formulate the main equations describing the dynamics of an ideal fluid and consider some particular cases, such as the static limit, stationary motion and the equation for sound waves. Of course, our consideration will be very far from being complete, and hence the reader is advised to consult advanced books, such as [6, 13, 22], to learn more on the subject.

In case the reader meets difficulties with vector calculus, we recommend looking at the Appendix A before reading this section.

10.2 Pascal's Law

The fluid is always subject to external and internal forces which can be characterized by pressure. If we consider a small oriented element of surface $d\mathbf{S} = \hat{\mathbf{n}}dS$, immersed in the fluid (see Fig. 10.1), then the force acting on this element is proportional to the area dS,

$$d\mathbf{F} = Pd\mathbf{S} = P\hat{\mathbf{n}}dS. \tag{10.1}$$

The coefficient P is called pressure. The Eq. (10.1) means that there are no tangent stresses in the fluid, and that the force acting on $d\mathbf{S}$ has its direction parallel to the normal unit vector $\hat{\mathbf{n}}$. Experiments confirm this feature and also tell us that the pressure P in the given point does not depend on the orientation of the surface $d\mathbf{S}$. In other words, the pressure in the given point of the fluid is the same in all directions. This statement is known as Pascal's law.

The Pascal's law can be easily applied to calculate the pressure of a homogeneous incompressible fluid in the gravitational field $\mathbf{g} = -g\hat{\mathbf{k}}$. Consider the vertical cylindric column of fluid of density ρ, area of the base S and depth h, as shown in Fig. 10.2. We assume that the pressure on the top of the column is the atmospheric one, P_A and the purpose is to evaluate the pressure at the bottom of the column. For this end we can use Newton's Second Law for the column, assuming that the the column is in rest. The pressure forces acting on the lateral sides cancel due to the symmetry. Taking into account the two pressure forces and the gravitational force acting on the mass of the column ρSh, we arrive at the equation

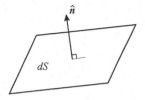

Fig. 10.1 A small oriented element of surface

$$- P_A S \hat{\mathbf{k}} - \rho S h g \, \hat{\mathbf{k}} + (P_A + P) S \hat{\mathbf{k}} = 0. \qquad (10.2)$$

Here the second term in the left hand side is the weight force of the column and the first and third terms are due to the pressure on the column from the top and bottom, respectively. The solution of (10.2) can be easily found, and we obtain the expression $P = \rho g h$ for the extra pressure at the depth h, compared to atmospheric pressure at the top. According to Pascal's law, this pressure does not depend on the orientation of the surface on which the corresponding force acts.

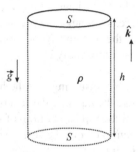

Fig. 10.2 Vertical cylindric fluid column of density ρ

Blaise Pascal, who invented the law of isotropic pressure, used it for a very illuminating experiment in 1646. A well-isolated and strong barrel was linked to the relatively thin tube of the large height h (see Fig. 10.3). When the barrel and the tube are filled by water, the pressure of the barrel increases by the magnitude of $\rho g h$. As a result, for a sufficiently large h, the barrel bursts, independent of the fact that the volume of the water in the tube is very small compared to the volume of the barrel. In the original Pascal experiment the tube was 10 m long and it was sufficient to achieve the desired effect.

Exercise

1. *a)* Discuss the formula (10.1) at a qualitative level. If we fix the direction of the unit vector $\hat{\mathbf{n}}$, from which side of the surface does the force act? The element of the surface is infinitely thin and therefore has zero mass. How it can be in equilibrium? Show that a tangential component added to Eq. (10.1) leads to a contradiction.

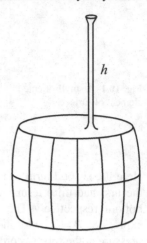

Fig. 10.3 Pascal's experiment: barrel linked to a thin
tube

b) Is it possible to have a negative pressure? In the affirmative case suggest an example. If we have a high vertical column of fluid with a piston on the top of it, which
creates a negative pressure on the top of the column, can the pressure be positive
at the bottom? What happens at the point of zero pressure? Can this point be found
experimentally?

2. *a)* Assuming that the height of barrel in the Pascal's experiment was 1 m, calculate the numerical effect of the extra pressure created by water in the 10 m-long tube
on the top and bottom of the barrel. Calculate the extra force acting on each square
centimeter of the surface of the barrel and compare it to atmospheric pressure.
b) Imagine that the same barrel is completely isolated and has, inside it, air under
atmospheric pressure. What is the depth in the sea at which it will be broken?
c) The barrel in the original experiment was made of wood. Assuming that the experiment is repeated with another material which is five times stronger, what is the
minimal necessary length of the thin tube with water?

3. An open cylindrical barrel with water is rotating around its symmetry axis with
constant angular velocity ω. Using the Pascal law and the Einstein's equivalence
principle, establish the equation describing the geometry of the surface of the
rotating water.[1]

Hint and answer. According to the Pascal's law, the pressure on the surface of the
rotating fluid is equal to atmospheric pressure. Then, according to the equivalence
principle, the normal vector to the surface is parallel to the effective gravitational
acceleration,

$$\mathbf{g}_{eff} = \mathbf{g} - \mathbf{a}_c = -g\hat{\mathbf{k}} - r\omega^2\hat{\mathbf{r}},$$

---

[1] This problem was already considered in Chap. 3, but here we treat it in a little bit different way.

in cylindrical coordinates. Starting from this expression one can easily arrive at the result. The equation describing the surface of the rotating fluid is $z = \omega^2 r^2/2g$ in cylindrical coordinates.

4. Starting from Pascal's law, derive Archimede's law.

Hint. There may be two possible solutions. One can consider a body as a sum of thin vertical cylinders, each of them with two horizontal surfaces as bases. Alternatively, one can write an overall force of pressure acting on the surface of the body as a surface integral and transform it to a volume integral by using the Gauss-Ostrogradsky theorem. Try to use both methods and show they provide equivalent results.

10.3 Continuity Equation

Consider a small body (V) with volume V which is part of the fluid. We assume that (V) is bounded by the surface (∂V). The fluid can pass freely through the boundary (∂V) and hence the mass of the fluid inside (V) can change with time. Let us construct the equation which describes this dependence and comes from mass conservation. At any instant of time, the mass of fluid inside (V) is given by the volume integral of its density, $\int_{(V)} \rho dV$. Therefore, the variation of mass in the volume (V) during a small time interval Δt is [2]

$$\Delta m = \Delta t \cdot \frac{d}{dt} \int_{(V)} \rho dV. \tag{10.3}$$

Obviously, this change of mass is due to the flux through the surface (∂V). The mass of the fluid which passes via the element of the surface $d\mathbf{S} = \hat{n} dS$ per unit of time is given by the scalar product $\rho \mathbf{v} \cdot d\mathbf{S}$. The sign of this expression depends on the orientation of the normal vector $\hat{n}$. For a closed simple surface such as (∂V) we assume by default that the normal vector $\hat{n}$ points to the outside of the volume (V). The total mass of the fluid which passes through the surface of the boundary to the exterior of (∂V), in the time interval Δt, is given by the surface integral of the second type,

$$\Delta m = -\Delta t \cdot \oint_{(\partial V)} \rho \mathbf{v} \cdot d\mathbf{S} \tag{10.4}$$

and, obviously, it is equal to the change of mass inside (V). In this way we arrive at the equation which describes the conservation of mass per unit of time,

---

[2] In this chapter we will denote all multiple integrals by a unique symbol, as is common in Physics literature. For example, a volume integral is $\int_{(V)} \rho dV$ instead of $\iiint_{(V)} \rho dV$.

$$\frac{d}{dt}\int_{(V)} \rho\, dV = -\oint_{(\partial V)} \rho \mathbf{v}\cdot d\mathbf{S} = -\int_{(V)} \operatorname{div}(\rho \mathbf{v})\, dV. \qquad (10.5)$$

In the last equality we used the Gauss-Ostrogradsky law. One can rewrite equation (10.5) in the form

$$\int_{(V)} \left[\frac{\partial \rho}{\partial t} + \operatorname{div}(\rho \mathbf{v})\right] dV = 0. \qquad (10.6)$$

Let us note that the time derivative of ρ is partial here, because ρ depends also on space coordinates. Since the last equation must hold for an arbitrary (V), we arrive at the local equation[3]

$$\frac{\partial \rho}{\partial t} + \operatorname{div}(\rho \mathbf{v}) = 0. \qquad (10.7)$$

This is called the continuity equation for a fluid. It is sometimes useful to introduce a new vector called density of current of the fluid, $\mathbf{j} = \rho \mathbf{v}$. Then the last equation can be cast into the form $\dot{\rho} + \operatorname{div}\mathbf{j} = 0$, where we denote partial time derivative with a dot.

Another interesting equivalent form of Eq. (10.7) is

$$\frac{\partial \rho}{\partial t} + \rho \operatorname{div}\mathbf{v} + \mathbf{v}\cdot \operatorname{grad}\rho = 0. \qquad (10.8)$$

Exercise 1.

(a) Derive (10.8) starting from (10.6).
(b) Suggest a physical interpretation for the condition $\operatorname{div}\mathbf{v} = 0$, namely link it to incompressibility of the fluid.
(c) In the case when the condition $\operatorname{div}\mathbf{v} = 0$ is satisfied, use Eq. (10.8) and suggest physical arguments explaining why $\mathbf{v}\cdot \operatorname{grad}\rho = 0$ implies $\dot{\rho} = 0$.

Hint. For the second task, consider the integral of $\operatorname{div}\mathbf{v}$ over (V). In the third case, derive and then use the following expression for the total derivative of the density:

$$\frac{d\rho}{dt} = \frac{\partial \rho}{\partial t} + \mathbf{v}\cdot \operatorname{grad}\rho. \qquad (10.9)$$

The detailed discussion of equivalent relation will be given in the next section.

---

[3] The term "local" means all quantities in the given expression correspond to the same point in space. For example, Eq. (10.5) does not belong to this category and Eq. (10.6) does.

10.4 Euler's Equation

The next step is to deduce the analog of Newton's second law for the ideal fluid. For this purpose we have to consider an infinitesimal element of the fluid, for example a one in some compact and very small volume (V). This volume is moving according to the fluid flow and since (V) is so small, we can attribute to it the velocity $\mathbf{v}$. At any instant of time it has the mass

$$m = \int_{(V)} \rho dV \approx \rho V, \qquad (10.10)$$

where we assumed that the continuous density function ρ does not really change within the volume V. The quality of this approximation becomes better with smaller compact volume V, and in what follows we assume that this volume is so small that the approximation is perfect. At any instant, the momentum of the fluid in the volume V is given by $\mathbf{p} = m\mathbf{v}$. According to Newton's second law, the force acting on the fluid in V is equal to

$$\mathbf{F} = m\frac{d\mathbf{v}}{dt} = \rho V \frac{d\mathbf{v}}{dt}. \qquad (10.11)$$

In order to use this equation we need to rewrite the complete time derivative in a more detailed form. Remember that the volume V is moving according to the stream of the fluid. Therefore the overall dependence of time for any quantity, such as $\mathbf{v}$ (or ρ in the Exercise at the end of the previous section) can be split into direct time dependence covered by the partial derivative $\partial \mathbf{v}/\partial t$ and the dependence on coordinates x, y, z, each of them being a function of time. Using the chain rule we arrive at the following relation:

$$\frac{d\mathbf{v}}{dt} = \frac{\partial \mathbf{v}}{\partial t} + \frac{\partial \mathbf{v}}{\partial x}\frac{dx}{dt} + \frac{\partial \mathbf{v}}{\partial y}\frac{dy}{dt} + \frac{\partial \mathbf{v}}{\partial z}\frac{dz}{dt}, \qquad (10.12)$$

The last three terms can be presented in a more useful form using vector notation, and we get

$$\frac{d\mathbf{v}}{dt} = \frac{\partial \mathbf{v}}{\partial t} + \frac{\partial \mathbf{v}}{\partial \mathbf{r}}\frac{d\mathbf{r}}{dt} = \frac{\partial \mathbf{v}}{\partial t} + (\mathbf{v}, \nabla)\mathbf{v}. \qquad (10.13)$$

The last two forms of the force deserve an extra comment. Each space point of the fluid is characterized by its own velocity, hence, at any instant of time, $\mathbf{v} = \mathbf{v}(\mathbf{r})$. The notation $\partial \mathbf{v}/\partial \mathbf{r}$ means gradient,

$$\frac{\partial \mathbf{v}}{\partial \mathbf{r}} = \hat{\mathbf{i}}\frac{\partial \mathbf{v}}{\partial x} + \hat{\mathbf{j}}\frac{\partial \mathbf{v}}{\partial y} + \hat{\mathbf{k}}\frac{\partial \mathbf{v}}{\partial z} \qquad (10.14)$$

and hence

$$\frac{\partial \mathbf{v}}{\partial \mathbf{r}} = \left(v_x \frac{\partial}{\partial x} + v_y \frac{\partial}{\partial y} + v_z \frac{\partial}{\partial z} \right) \mathbf{v} = (\mathbf{v}, \nabla) \mathbf{v}. \tag{10.15}$$

Let us now consider the right hand side of Newton's law. The motion of the element (V) of the fluid occurs in the gravitational field $\mathbf{g}$ and therefore we have the corresponding force $\mathbf{F}_g = \rho V \mathbf{g}$. On top of that, the element of the fluid is subject to the force of pressure. If pressure is constant everywhere, this force is zero, of course. Let us now imagine that pressure is position-dependent, $P = P(\mathbf{r})$. Obviously, the force of pressure acts from the side where the pressure is higher to the one where it is lower. For a small volume V we arrive at the expression (see also Exercise 2) $\mathbf{F}_{press} = -V \operatorname{grad} P$.

Finally, we can put all elements together and remember that the mass m is proportional to the volume V,

$$m = \rho V.$$

The dynamical equation can be attributed to the unit volume, and taking into account Eq. (10.13) it can be cast into the form

$$\rho \frac{\partial \mathbf{v}}{\partial t} + \rho (\mathbf{v}, \nabla) \mathbf{v} = \rho \mathbf{g} - \operatorname{grad} P. \tag{10.16}$$

Dividing by ρ, we can rewrite (10.16) in the form which is considered the standard one for the Euler equation for the ideal fluid,

$$\frac{\partial \mathbf{v}}{\partial t} + (\mathbf{v}, \nabla) \mathbf{v} = \mathbf{g} - \frac{\operatorname{grad} P}{\rho}. \tag{10.17}$$

This equation can be seen as a version of Newton's Second Law, for it describes the dynamics of a unit volume of an ideal fluid moving in a gravitational field. As a simple particular application of the Eq. (10.17), let us consider a static fluid in the homogeneous gravitational force of the Earth, $\mathbf{g} = -g\hat{\mathbf{k}}$. In this case we have $\mathbf{v} = 0$ and therefore the Euler equation boils down to

$$\operatorname{grad} P = \mathbf{g}\rho = -g\rho\hat{\mathbf{k}}. \tag{10.18}$$

In this way we immediately arrive at Pascal's law, $P = P_0 - \rho g z$. Here the initial value P_0 corresponds to atmospheric pressure at the surface of the fluid at $z = 0$.

Exercise 2. *(a)* A more rigorous derivation of the pressure force is as follows: consider the force of pressure acting on the volume (V) in form of a surface integral over the boundary of the volume,

$$\mathbf{F}_{press} = - \oint_{(\partial V)} P(\mathbf{r}) \cdot d\mathbf{S}, \tag{10.19}$$

where the negative sign is due to the fact that $dS = \hat{n}dS$, where $\hat{n}dS$ is the *external* normal vector to the surface. Using the Gauss-Ostrogradsky theorem, reduce this expression to the volume integral over (V) and show that the expression $\mathbf{F}_{press} = -V \operatorname{grad} P$ is the result for a very small volume.

(b) Use the same method as in point (a) to derive the whole equation (10.17) in a more complete mathematical way.

Exercise 3. Consider a spherically symmetric static star made of an ideal fluid with constant density ρ. The radius of the star is R. Calculate the pressure of the fluid at the distance $r \leq R$ from the center of the star.

Hint: The solution can be done by using Eq. (10.17) and directly comparing the gravitational force acting on the layer of the fluid between r and R and the force of pressure at the level r. Use both methods and compare the results.

$$\textbf{Answer:} \qquad P = \frac{2\pi G \rho^2}{3}\left(R^2 - r^2\right).$$

Exercise 4. *(a)* Show that a fluid which is static under the action of the gravitational force satisfies the equation

$$\operatorname{grad} P(\mathbf{r}) = -\rho \operatorname{grad} \Phi(\mathbf{r}), \qquad\qquad (10.20)$$

where Φ is the gravitational potential per unit of mass, $\mathbf{g} = -\operatorname{grad} \Phi$.

(b) Use Eq. (10.20) and Einstein's equivalence principle to calculate the pressure at all points of an open cylindric barrel with water, which is rotating around its symmetry axis with a constant angular velocity ω, in a constant gravitational field $\mathbf{g} = -g\hat{\mathbf{k}}$. The radius of the barrel is R and the volume of the fluid is V.

Solution. From Exercise 10.2 of Sect. 10.2 we know that the form of the rotating surface in cylindric coordinates is $h = \omega^2 r^2 / 2g + H$, where $h(r)$ is the coordinate z of the surface at the point r and $H = h(0)$. It is easy to calculate

$$H = \frac{V}{\pi R^2} - \frac{\omega^2 R}{4g}.$$

Obviously, the pressure at the point with coordinates r, φ, z inside the barrel is defined by the expression

$$P = \rho g(h - z).$$

Using this relation one can obtain the result

$$P = \rho g \left[\frac{V}{\pi R^2} - \frac{\omega^2 R^2}{4g} + \frac{\omega^2 r^2}{2g} - z\right]. \qquad\qquad (10.21)$$

10.5 Bernoulli's Equation

In this section we consider a particular case of Euler's equation, namely the steady flow of the incompressible fluid. Another particular case, concerning sound waves propagating in a homogeneous fluid, will be dealt with in the next section. Of course, there are a number of other very important applications, but we selected these two, because they are sufficiently simple and very illustrative.

Bernoulli's equation describes a steady flow of the fluid. The term "steady" means that at each point of the space $\mathbf{r}$, the vector of velocity $\mathbf{v}$ does not change with time. So, $\mathbf{v}$ depends on $\mathbf{r}$, but not on t. In this case we can introduce the notion of a streamline, or line of flux. For simplicity, one can think that a such a line is reproduced by a very light particle of dust which follows the stream starting from a certain initial point. Mathematically, the flux line can be parameterized by a natural parameter l along the curve, such that the unit vector $\hat{\mathbf{v}} = \mathbf{v}/v$ is a tangent vector to this curve.

Indeed, the flux consists of infinitely many streamlines. If we restrict our attention to one selected streamline, the dynamical equation can be very much simplified. Our first purpose will be to find the dependence of the absolute value of velocity of the fluid along the streamline (no information about direction of this vector will be available in this case, of course) on the pressure and position of the given point.

As a first step, we rewrite the Euler equation (10.17) by using the mathematical identity

$$(\mathbf{v}, \nabla)\mathbf{v} = \frac{1}{2}\,\mathrm{grad}\,v^2 - [\mathbf{v}, \mathrm{rot}\,\mathbf{v}] . \tag{10.22}$$

Replacing this relation into (10.17), after very simple algebra we arrive at the equation

$$\frac{\partial \mathbf{v}}{\partial t} - [\mathbf{v}, \mathrm{rot}\,\mathbf{v}] = -\,\mathrm{grad}\left(\frac{P}{\rho} + \frac{v^2}{2} + \Phi\right), \tag{10.23}$$

where we reintroduced the notation Φ for the Newtonian gravitational potential per unit mass, $\mathbf{g} = -\,\mathrm{grad}\,\Phi$. In this case we consider, for the sake of simplicity, an incompressible fluid, which has a constant density ρ.

The next step is to multiply Eq. (10.23) by the unit vector $\hat{\mathbf{v}}$ and remember that for steady flow

$$\hat{\mathbf{v}} \cdot \dot{\mathbf{v}} = \frac{1}{2v}\frac{dv^2}{dt} = 0 .$$

Also, $\hat{\mathbf{v}} \cdot [\mathbf{v}, \mathrm{rot}\,\mathbf{v}] = 0$ and the product of $\hat{\mathbf{v}}$ with the gradient of some scalar function gives the derivative of this function with respect to the natural parameter l along the streamline. Then we arrive at the equation

$$\frac{d}{dl}\left(\frac{P}{\rho} + \frac{v^2}{2} + \Phi\right) = 0, \tag{10.24}$$

and hence

$$\frac{\rho v^2}{2} + P + \rho \Phi = 0, \tag{10.25}$$

which is called the Bernoulli's equation. One can easily recognize that the expression $\rho v^2/2$ in the last equation is nothing but the kinetic energy per unit volume. Furthermore, it is obvious that the term $-dP/dl$ has the form of a potential force acting along the streamline. One can see that the Eq. (10.25) is a kind of energy conservation law for the small element of the fluid moving along the streamline. In case of constant homogeneous gravity $\mathbf{g} = -g\hat{\mathbf{k}}$, the gravitational potential in (10.25) has the form $\Phi = gz$.

The use of Bernoulli's law (10.25) is restricted to the case of a frictionless fluid with constant density, but it can be still very useful in many practical applications. Let us consider only two examples.

Example 1. In the case when the flow of the fluid occurs (we assume there is no turbulence) in a horizontal tube with cross-sectional area that depends on the point along the tube, $S = S(l)$, the continuity equation gives

$$v(l) = v(0) \frac{S(0)}{S(l)}. \tag{10.26}$$

Then Bernoulli's equation (10.25) tells us that the pressure depends on the natural parameter l along the tube as

$$P(l) = P_0 + \frac{\rho v^2(0)}{2} \left[1 - \frac{S^2(0)}{S^2(l)} \right]. \tag{10.27}$$

It is easy to see that in those parts of the tube which are relatively narrow, the velocity is greater according to (10.26) and the pressure is smaller according to (10.27).

Example 2. Consider the case when water falls freely in the air after it leaves the tap. In this case the pressure along the flow is always equal to atmospheric pressure and therefore it does not depend on the velocity of the fluid. Then Bernoulli's equation (10.25) tells us that the speed of the water follows the free fall law

$$v(h) = \sqrt{v_0^2 + 2gh}, \tag{10.28}$$

where h is the distance to the tap and v_0 is the initial speed of the water. The continuity equation tells us that the cross-sectional area of the flow depends on h as

$$S(h) = \frac{S_0}{\sqrt{1 + 2gh/v_0^2}}, \tag{10.29}$$

Exercise 5. Deduce the Eq. (10.26) from the continuity equation (10.7).

Exercise 6.

(a) Verify the formula for the pressure of the rotating fluid (10.21) using Bernoulli's equation.

(b) Use Bernoulli's equation to discuss pressure at different points of the flux of an ideal fluid moving in the horizontal tube which makes a curve with the curvature radius R. Assume that the diameter of the tube is $d \ll R/2$. Can all points of the fluid in the transverse section (orthogonal to the velocity of the flux) have the same velocity?

10.6 Sound Wave Equation

There are different sorts of waves in a fluid, and sound waves propagating in the homogeneous fluid at rest is a very interesting example. One of the reasons is that the equation for this kind of wave is in some respects similar (and in others not so) to electromagnetic waves. One of the reasons why we choose this particular example is that, in our opinion, it is important that the student can appreciate this analogy. Also, acoustic waves have many important applications, from technology to cosmology.

Consider a fluid which has a very small deviation from a static equilibrium. The background equilibrium state of the fluid is characterized by the density ρ_0 and pressure P_0 in each point of the fluid. We assume that these two quantities are related by some equation of state, such that $P = P(\rho)$ and, in particular, $P_0 = P(\rho_0)$. Let us note that the consideration that follows below requires that the same equation of state $P = P(\rho)$ is valid for both background and perturbations.

The sound wave is a small perturbation of density and pressure of the fluid, such that

$$\rho = \rho_0 + \delta\rho, \qquad P = P_0 + \delta P, \tag{10.30}$$

which can propagate from one point of the space to another. Our intention is to consider only small perturbations $\delta\rho$ and δP. The motion of the fluid in each point is also characterized by the velocity $\mathbf{v}$. As far as the background fluid is at rest, the velocities $\mathbf{v} = \mathbf{v}(t, \mathbf{r})$ are only due to the perturbations. Therefore, they should be as small as $\delta\rho$ and δP. As usual, the notion of "small" can be defined only by comparison with some other quantity, but it is better to postpone this part of the consideration to the moment when we obtain the solution for the perturbations and will be able to analyze them (see Exercise 5). In practice, we will assume that $\delta\rho$, δP and $\mathbf{v}$ are so small that we can restrict consideration to linear order in these quantities.

Our purpose is to obtain the dynamical equations for the small perturbations $\delta\rho$, δP and $\mathbf{v}$. To this end we shall use the continuity equation (10.7),

$$\frac{\partial\rho}{\partial t} + \text{div}\,(\rho\mathbf{v}) = 0 \tag{10.31}$$

and the Euler equation (10.17),

$$\frac{\partial \mathbf{v}}{\partial t} + (\mathbf{v}, \nabla)\,\mathbf{v} = \mathbf{g} - \frac{\operatorname{grad} P}{\rho}, \tag{10.32}$$

with all quantities understood as sums of background part and perturbations, according to (10.30). Let us start from the Euler equation. Replacing (10.30) into (10.32) we note that the terms in the right hand side of this equation cancel each other for the static background and we get, in the first order in $\delta\rho$ and δP,

$$\frac{\partial \mathbf{v}}{\partial t} + (\mathbf{v}, \nabla)\,\mathbf{v} = -\frac{\operatorname{grad} \delta P}{\rho_0}. \tag{10.33}$$

The conclusion is that, for small perturbations over the static background, we can disregard the gravitational force. Furthermore, assuming that the perturbations are small, we admit that the velocities are also small (otherwise the perturbations become not so small in a short time). Therefore, the term quadratic in velocities, $(\mathbf{v}, \nabla)\,\mathbf{v}$, can be neglected compared to other terms and we arrive at the truncated Euler equation for the perturbations

$$\frac{\partial \mathbf{v}}{\partial t} = -\frac{\operatorname{grad} \delta P}{\rho_0}. \tag{10.34}$$

As it was already discussed above, our fluid satisfies a certain equation of state $P = P(\rho)$, which corresponds to the situation of an ideal fluid, hence the motion occurs without any kind of friction. Then we can replace

$$\delta P = \left(\frac{\partial P}{\partial \rho}\right)_s \delta\rho, \tag{10.35}$$

where the index s indicates that the entropy is constant (this simply means no heat is produced in the process). Then the Eq. (10.34) can be cast into the form

$$\frac{\partial \mathbf{v}}{\partial t} = -\frac{1}{\rho_0}\left(\frac{\partial P}{\partial \rho}\right)_s \operatorname{grad} \delta\rho. \tag{10.36}$$

The Eq. (10.36) follows from Euler's equation. Now we consider the continuity equation and replace (10.30) into Eq. (10.31). Taking into account that this equation is satisfied for P_0 and ρ_0 with zero velocity, for the linear perturbations we obtain the equation

$$\frac{\partial \delta\rho}{\partial t} = -\rho_0 \operatorname{div}\mathbf{v}. \tag{10.37}$$

The two Eqs. (10.36) and (10.37) have all we need to describe an acoustic wave in the ideal fluid, in fact we only need to combine them in the right way. Taking a partial derivative $\partial/\partial t$ of (10.37) and replacing there (10.36), we arrive at the equation

$$\frac{\partial^2 \delta\rho}{\partial t^2} = \rho_0 \cdot \operatorname{div}\left[\frac{1}{\rho_0}\left(\frac{\partial P}{\partial \rho}\right)_s \cdot \operatorname{grad}\delta\rho\right] = c^2 \Delta\,\delta\rho, \qquad (10.38)$$

where we introduced the notation $c^2 = (\partial P/\partial \rho)_s$, used the relation $\operatorname{div}(\operatorname{grad}\varphi) = \Delta\varphi$ and assumed that the space derivatives of the background density ρ_0 are negligible compared to the space derivatives of the perturbation $\delta\rho$.

The Eq. (10.38),

$$\left(\frac{\partial^2}{\partial t^2} - c^2\Delta\right)\delta\rho = 0,$$

describes free wave which propagates with the velocity c. In order to see this, we can consider wave propagating in the direction x,

$$\frac{\partial^2 \delta\rho}{\partial t^2} + c^2\frac{\partial^2 \delta\rho}{\partial x^2} = 0, \qquad (10.39)$$

and rewrite this equation in the form

$$\left(\frac{\partial}{\partial t} + c\frac{\partial}{\partial x}\right)\left(\frac{\partial}{\partial t} - c\frac{\partial}{\partial x}\right)\delta\rho = 0. \qquad (10.40)$$

Looking at this equation and taking into account that the two parenthesis do commute with each other, one can see that the general solution of this equation has a form, which depends on two arbitrary functions f_1 and f_2

$$\delta\rho(x,t) = f_1(ct - x) + f_2(ct + x), \qquad (10.41)$$

Indeed, the fact that this solution is general, requires the use of some known theorems of Mathematics. An alternative derivation of (10.41) is postponed to the exercises.

The two parts of the solution (10.41) correspond to a plane sound waves propagating forward and backward along the axes x. One can easily obtain similar equations and solutions for waves propagating along y and z.

Exercises.

1. *(a)* Derive the solution (10.41) in a more detailed way. For this purpose one has to introduce the new variables $\xi = x - ct$ and $\eta = x + ct$ and show that (10.40) transforms into

$$\frac{\partial^2 \delta\rho}{\partial \xi\,\partial \eta} = 0. \qquad (10.42)$$

The general solution of this equation is a sum of two functions, $f_1(\xi)$ and $f_1(\eta)$, that directly gives us (10.41).

(b) Write down the solutions for the waves propagating along y and z and the solution for the wave propagating along the direction of the vector $\hat{n} = \hat{i}\cos\alpha + \hat{j}\cos\beta + \hat{k}\cos\gamma$. What is the form of the equation for this solution? How does one define the initial conditions for Eq. (10.38)?

Answer: $\delta\rho(t,\mathbf{r}) = f_1(ct - \hat{n}\cdot[\mathbf{r} - \mathbf{r}_0]) + f_2(ct + \hat{n}\cdot[\mathbf{r} - \mathbf{r}_0])$ (10.43)

and the initial conditions should be defined such that this form of dependence on t and $\mathbf{r}$ are maintained.

2. (a) Using the result of the previous exercise, write down the expression for $\mathbf{v}(t,\mathbf{r})$ for the wave propagating in the direction $\hat{n}$. Show that $\hat{v} = \mathbf{v}/v = \pm\hat{n}$. Evaluate the ratio v/c and show that when it is small, the linear approximation used to derive Eq. (10.34) is justified.
(b) Discuss whether, in the general case, one can assume that $\mathbf{v} = \mathrm{grad}\,\Psi$, where Ψ is a scalar function of space coordinates and time $\Psi = \Psi(t,\mathbf{r})$.

Solution. It is easy to verify that, according to (10.43),

$$\mathrm{grad}\left[\delta\rho(t,\mathbf{r})\right] = \frac{\hat{n}}{c}\left[\frac{\partial f_2}{\partial t} - \frac{\partial f_1}{\partial t}\right].$$ (10.44)

Replacing this relation into Eq. (10.36) we arrive at the equation

$$\frac{\partial \mathbf{v}}{\partial t} = \frac{\hat{n}}{c}\frac{\partial(f_1 - f_2)}{\partial t},$$

that immediately gives us

$$\mathbf{v}(t,\mathbf{r}) = \frac{\hat{n}}{\rho_0}\left[f_1(ct - \hat{n}\cdot[\mathbf{r} - \mathbf{r}_0]) + f_2(ct + \hat{n}\cdot[\mathbf{r} - \mathbf{r}_0])\right].$$ (10.45)

Obviously, $\hat{v} = \pm\hat{n}$ in this case.

Observation. Waves of this type, when the perturbation velocities are parallel to the direction of propagation of the wave, are called longitudinal. There are other types of waves, called transverse, when the velocities of perturbations are orthogonal to the direction of propagation of the wave. In the case of hydrodynamics, the transverse waves are, for instance, the waves of gravitational origin on the surface of water. Many of those can be observed at the seaside, for example at Copacabana in Rio de Janeiro. Electromagnetic waves in vacuum are also transverse.

3. Use Eq. (10.36) to find the behavior of the velocity $\mathbf{v}$ which follows from (10.41). Consider an outgoing wave with $f_1(x - ct) = A\cos(\omega t - kx)$ with $k = \omega/c$. Discuss the limits of the linear approximation for different values of the frequency ω. Show that this approximation is indeed equivalent to the condition $v \ll c$.

Solution. As we already know from the previous exercise, in this case $\hat{\mathbf{v}} = \pm\hat{\mathbf{i}}$, so we do not need vector notation. Using Eq. (10.32) we obtain

$$\frac{\partial v}{\partial t} = -\frac{c^2 A}{\rho_0} \cos(\omega t - kx). \qquad (10.46)$$

Integrating this and ignoring the constant term (remember the background solution is static), we arrive at

$$v = -\frac{c^2 A}{\rho_0 \omega} \sin(\omega t - kx). \qquad (10.47)$$

Now, calculating $(\nabla, \mathbf{v})\mathbf{v} = \hat{\mathbf{i}} \cdot dv/dx$ and comparing it to Eq. (10.46), we arrive at the conclusion that

$$(\nabla, \mathbf{v})\mathbf{v} = -\frac{v}{c} \frac{\partial v}{\partial t} \hat{\mathbf{i}}. \qquad (10.48)$$

Observation. One can use an expansion into Fourier series to argue that the condition $v \ll c$ corresponds to the linear approximation for a general continuous function f_1.

4. (a) Write down the equation for the sound wave in spherical coordinates. Consider $\delta\rho = \varphi$ and use the change of variables $\varphi = \chi(r)/r$. Show that the amplitude of the outgoing spherical wave $\varphi = \varphi(r - ct)$ emitted at the beginning of the coordinate system behaves proportionally to $1/r$.

Solution. The problem has spherical symmetry, therefore φ can depend only on the radius r and not on the angle variables. Then the Laplace operator in spherical coordinates gets simplified and the whole equation boils down to

$$c^2 \frac{\partial^2 \varphi}{\partial t^2} - \Delta\varphi = c^2 \frac{\partial^2 \varphi}{\partial t^2} - \frac{1}{r^2} \frac{\partial}{\partial r}\left(r^2 \frac{\partial\varphi}{\partial r}\right) = 0. \qquad (10.49)$$

Simple algebra shows that $\chi(r)$ satisfies the equation for the plane wave

$$c^2 \frac{\partial^2 \chi}{\partial t^2} - \frac{\partial^2 \chi}{\partial r^2} = 0, \qquad (10.50)$$

and, therefore, the amplitude of the field χ does not change. As a result, for the propagating field φ we have $\varphi \sim 1/r$.

(b) Solve the same problem for the sound emitted by a linear infinite line source, $x = y = 0$. Show that the amplitude of the cylindrically symmetric wave has the coordinate dependence which is different from the one in the spherically symmetric case.

Hint and answer. One has to use cylindric coordinates. In this case $\varphi \sim 1/\sqrt{r}$.

5. What will change in the wave equation if we do not assume that $\operatorname{grad}\rho_0$ is negligible in Eq. (10.38)? Explore the corresponding equation for the wave propagating in the direction OX. Does the solution look like the one for the *ideal* fluid? Consider a simple single-frequency solution from Exercise 3. The quantity $\lambda = 2\pi c/\omega$ is called the wavelength of the acoustic wave. Discuss the relation between the typical size of inhomogeneity of the background solution ρ_0 and λ in relation to the approximation which we were using above.

Appendix A
Fundamentals of Vector Analysis

Abstract The purpose of this appendix is to present a consistent but brief introduction to vector calculus. For the sake of completeness, we shall begin with a brief review of vector algebra. It should be emphasized that this appendix cannot be seen as a textbook on vector algebra and analysis. In order to learn the subject in a systematic way, the reader can use special textbooks. At the same time, we will consider here a content which is supposed to be sufficient for applications in Classical Mechanics, at the level used in this book.

A.1 Vector Algebra

In this paragraph, we use a constant and global basis which consists of orthonormal vectors $\hat{\mathbf{i}}, \hat{\mathbf{j}}, \hat{\mathbf{k}}$. A vector is represented geometrically by an oriented segment (arrow), which is characterized by length (also called absolute value, or modulus, or magnitude of a vector) and direction. Any vector $\mathbf{a}$ can be expressed as a linear combination of the basis vectors,

$$\mathbf{a} = a_1\hat{\mathbf{i}} + a_2\hat{\mathbf{j}} + a_3\hat{\mathbf{k}}. \tag{A.1}$$

The linear operations on vectors include:

(i) The multiplication by a constant k, which is equivalent to the multiplication of all components of a vector by the same constant:

$$k\mathbf{a} = (ka_1)\hat{\mathbf{i}} + (ka_2)\hat{\mathbf{j}} + (ka_3)\hat{\mathbf{k}}. \tag{A.2}$$

(ii) The sum of two vectors, $\mathbf{a}$ and $\mathbf{b}$, obtained by the addition operation is a vector with components equal to the sum of the components of the original vectors,

$$\mathbf{a} + \mathbf{b} = (a_1 + b_1)\hat{\mathbf{i}} + (a_2 + b_2)\hat{\mathbf{j}} + (a_3 + b_3)\hat{\mathbf{k}}. \tag{A.3}$$

I.L. Shapiro and G. de Berredo-Peixoto, *Lecture Notes on Newtonian Mechanics*,
Undergraduate Lecture Notes in Physics, DOI 10.1007/978-1-4614-7825-6,
© Springer Science+Business Media, LLC 2013

Besides the linear operations, we define the following two types of multiplication. The scalar product (dot product) between the two vectors, **a** and **b**, is defined as

$$\mathbf{a}\cdot\mathbf{b} = (\mathbf{a},\mathbf{b}) = ab\cos\varphi,\tag{A.4}$$

where a and b represent absolute values of the vectors **a** and **b**, given by $a = |\mathbf{a}|$ and $b = |\mathbf{b}|$, and φ is the smallest angle between these vectors.

The main properties of the scalar product (A.4) are commutativity,

$$\mathbf{a}\cdot\mathbf{b} = \mathbf{b}\cdot\mathbf{a},$$

and linearity,

$$\mathbf{a}\cdot(\mathbf{b}_1+\mathbf{b}_2) = \mathbf{a}\cdot\mathbf{b}_1+\mathbf{a}\cdot\mathbf{b}_1.\tag{A.5}$$

As a result, we can use the representation (A.1) for both vectors **a** and **b**, in this way we arrive at the formula for the scalar product expressed in components,

$$\mathbf{a}\cdot\mathbf{b} = a_1b_1+a_2b_2+a_3b_3.\tag{A.6}$$

The definition of vector product (cross product) between the two vectors is a little bit more complicated. The vector **c** is the vector product of the vectors **a** and **b**, and is denoted as

$$\mathbf{c} = \mathbf{a}\times\mathbf{b} = [\mathbf{a},\mathbf{b}],\tag{A.7}$$

if the following three conditions are satisfied:

- The vector **c** is orthogonal to the other two vectors, i.e., $\mathbf{c}\perp\mathbf{a}$ and $\mathbf{c}\perp\mathbf{b}$.
- The modulus of the vector product is given by

$$c = |\mathbf{c}| = ab\sin\varphi,$$

where φ is the smallest angle between the vectors **a** and **b**.
- The orientation of the three vectors, **a**, **b**, and **c**, is right handed (or dextrorotatory). This expression, as discussed in Chap. 2, means the following. Imagine a rotation of **a** until its direction matches with the direction of **b**, by the smaller angle between them. The vector **c** belongs to the axis of rotation and the only question is how to choose the direction of this vector. This direction must be chosen such that, by looking at the positive direction **c**, the rotation is performed clockwise. A simple way to memorize this guidance is to remember about the motion of a corkscrew. When the corkscrew turns **a** up to **b**, it advances in the direction of **c**. Another useful rule is the right-hand rule, rather commonplace in the textbooks of Physics.

The main properties of the vector product are the antisymmetry,

$$\mathbf{a}\times\mathbf{b} = -\mathbf{b}\times\mathbf{a}\tag{A.8}$$

(or anticommutativity), and the linearity,

$$[\mathbf{a}, (\mathbf{b}_1 + \mathbf{b}_2)] = [\mathbf{a}, \mathbf{b}_1] + [\mathbf{a}, \mathbf{b}_2].$$ (A.9)

The vector product can be expressed as a determinant, namely,

$$\mathbf{a} \times \mathbf{b} = \begin{vmatrix} \hat{\mathbf{i}} & \hat{\mathbf{j}} & \hat{\mathbf{k}} \\ a_1 & a_2 & a_3 \\ b_1 & b_2 & b_3 \end{vmatrix} = \hat{\mathbf{i}} \begin{vmatrix} a_2 & a_3 \\ b_2 & b_3 \end{vmatrix} - \hat{\mathbf{j}} \begin{vmatrix} a_1 & a_3 \\ b_1 & b_3 \end{vmatrix} + \hat{\mathbf{k}} \begin{vmatrix} a_1 & a_2 \\ b_1 & b_2 \end{vmatrix}.$$ (A.10)

Some important relations involving vector and scalar products will be addressed in the form of exercises.

Exercises

1. Using the definition of vector product, check that

$$\hat{\mathbf{i}} \times \hat{\mathbf{j}} = \hat{\mathbf{k}}, \qquad \hat{\mathbf{k}} \times \hat{\mathbf{i}} = \hat{\mathbf{j}}, \qquad \hat{\mathbf{j}} \times \hat{\mathbf{k}} = \hat{\mathbf{i}}.$$ (A.11)

Show that the magnitude of the vector product, $|\mathbf{a} \times \mathbf{b}|$, is equal to the area of the parallelogram with vectors $\mathbf{a}$ and $\mathbf{b}$ as edges.

2. We can combine both types of multiplication and build the so-called mixed product, which involves three independent vectors $\mathbf{a}$, $\mathbf{b}$ and $\mathbf{c}$,

$$(\mathbf{a}, \mathbf{b}, \mathbf{c}) = (\mathbf{a}, [\mathbf{b}, \mathbf{c}]) = \begin{vmatrix} a_1 & a_2 & a_3 \\ b_1 & b_2 & b_3 \\ c_1 & c_2 & c_3 \end{vmatrix}.$$ (A.12)

(a) Verify the second equality of (A.10).
(b) Show that the mixed product has the following cyclic property:

$$(\mathbf{a}, \mathbf{b}, \mathbf{c}) = (\mathbf{c}, \mathbf{a}, \mathbf{b}) = (\mathbf{b}, \mathbf{c}, \mathbf{a}).$$

(c) Show that the modulus of the mixed product, $|(\mathbf{a}, \mathbf{b}, \mathbf{c})|$, is equal to the volume of the parallelepiped with the three vectors $\mathbf{a}$, $\mathbf{b}$, and $\mathbf{c}$ as edges.
(d) Show that the mixed product $(\mathbf{a}, \mathbf{b}, \mathbf{c})$ has a positive sign in the case when the three vectors $\mathbf{a}$, $\mathbf{b}$, $\mathbf{c}$ (in this order!) have dextrorotatory orientation.

3. Another, interesting for us, quantity is the double vector product,

$$[\mathbf{a}, [\mathbf{b}, \mathbf{c}]].$$ (A.13)

Derive the following relation

$$[\mathbf{a}, [\mathbf{b}, \mathbf{c}]] = \mathbf{b} (\mathbf{a}, \mathbf{c}) - \mathbf{c} (\mathbf{a}, \mathbf{b}).$$ (A.14)

This identity is useful in many cases.

A.2 Scalar and Vector Fields

In the next paragraph we will consider differential operations performed on the scalar or vector fields. For this reason, here we introduce the notion of a field, including scalar and vector cases.

The scalar field is a function $f(\mathbf{r})$ of a point in space. Each point of the space M is associated with a real number, regardless of how we parameterize the space, i.e., regardless of the choice of a system of coordinates. As examples of scalar fields in physics, we can mention the pressure of air or its temperature at a given point.

In practice, we used coordinates to parametrize the space, and the scalar field becomes a function of the coordinates, $f(\mathbf{r}) = f(x,y,z)$. In order to have the property of coordinate-independence, mentioned above, the function $f(x,y,z)$ should obey the following condition: If we change the coordinates and consider, instead of x, y and z, some other coordinates, say x', y', and z', the form of the functional dependence $f(x,y,z)$ should be adjusted such that the value of this function in a given geometric point M would remain the same. This means that the new form of the function, denoted by f' is defined by the condition

$$f'(x',y',z') = f(x,y,z). \tag{A.15}$$

Example 1. Consider a scalar field defined on the coordinates x, y and z by the formula

$$f(x,y,z) = x^2 + y^2.$$

Find the shape of the field f' for other coordinates,

$$x' = x+1, \quad y' = z, \quad z' = -y. \tag{A.16}$$

Solution. To find $f'(x',y',z')$, we will use Eq. (A.15). The first step is to solve (A.16) with respect to the coordinates x,y,x and then just replace the solution in the formula for the field. We find

$$x = x'-1, \quad y = -z', \quad z = y' \tag{A.17}$$

hence

$$f'(x',y',z') = f\left(x(x',y',z'), y(x',y',z'), z(x',y',z')\right) = (x'-1)^2 + z'^2.$$

In addition to the coordinates, the scalar field may depend on the time variable, but since time and spatial coordinates are independent in Classical Mechanics, this appendix does not consider temporal dependence for scalar fields. The same is valid for other types of fields.

The next example of our interest is the vector field. The difference with the scalar field is that, in the vector case, each point of the space is associated with a vector, say, $\mathbf{A}(\mathbf{r})$. If we parameterize the points in space by their radius-vectors, we have

$$\mathbf{A} = \mathbf{A}(\mathbf{r}) = A_1(\mathbf{r})\hat{\mathbf{i}} + A_2(\mathbf{r})\hat{\mathbf{j}} + A_3(\mathbf{r})\hat{\mathbf{k}}, \tag{A.18}$$

where $A_{1,2,3}$ are the components of the vector field. If we use Cartesian coordinates x, y and z, the vector field becomes a set of three functions, written as

$$A_1(x,y,z), \quad A_2(x,y,z) \quad \text{and} \quad A_3(x,y,z).$$

A pertinent question is what is the law of transformation of these functions corresponding to the coordinate transformation, say, when one moves from x, y, z coordinates to the x', y', z' ones? We have already found an answer for scalar fields, but in the case of a vector the answer should be different from (A.15). The reason is that, in geometric terms, the components of an oriented segment may vary depending on the coordinate system. For example, for a constant vector, $\mathbf{a}$, we can always choose a new reference frame in such a way that only one of its components is different from zero.

The complete solution of the problem of transforming components of a vector will not be discussed here. Instead we suggest an interested reader to use some of the books dealing with vector and tensor analysis, e.g., in [2, 28, 29]. However, we can consider the answer in some particular cases of space transformations, especially related to rotations of the orthonormal basis.

In case when the basis remains orthogonal after transformation, the components of an arbitrary vector are transformed in the same way as the vector describing the position of a point, its radius-vector $\mathbf{r}$. For example, we consider a rotation angle φ in the plane XOY. Using Fig. A.1, one can find the law of transformation to the new coordinates, and the inverse transformation, in the form

$$\begin{aligned} x &= x'\cos\varphi - y'\sin\varphi & x' &= x\cos\varphi + y\sin\varphi \\ y &= x'\sin\varphi + y'\cos\varphi\,, & y' &= -x\sin\varphi + y\cos\varphi \\ z &= z' & z' &= z \end{aligned} \tag{A.19}$$

We can express these transformations in the matrix form

$$\begin{pmatrix} x \\ y \\ z \end{pmatrix} = (\Lambda) \begin{pmatrix} x' \\ y' \\ z' \end{pmatrix}, \qquad \begin{pmatrix} x' \\ y' \\ z' \end{pmatrix} = (\Lambda^{-1}) \begin{pmatrix} x \\ y \\ z \end{pmatrix}, \tag{A.20}$$

where

$$(\Lambda) = \begin{pmatrix} \cos\varphi & -\sin\varphi & 0 \\ \sin\varphi & \cos\varphi & 0 \\ 0 & 0 & 1 \end{pmatrix}$$

$$\text{and} \quad (\Lambda^{-1}) = \begin{pmatrix} \cos\varphi & \sin\varphi & 0 \\ -\sin\varphi & \cos\varphi & 0 \\ 0 & 0 & 1 \end{pmatrix}. \tag{A.21}$$

By definition, the law of transformation for the vector components is the same as for the coordinates, i.e.,

$$\begin{pmatrix} A_1 \\ A_2 \\ A_3 \end{pmatrix} = (\Lambda) \begin{pmatrix} A_1' \\ A_2' \\ A_3' \end{pmatrix}, \qquad \begin{pmatrix} A_1' \\ A_2' \\ A_3' \end{pmatrix} = (\Lambda^{-1}) \begin{pmatrix} A_1 \\ A_2 \\ A_3 \end{pmatrix}. \qquad (A.22)$$

The structure of formulas (A.22) has a general nature, i.e., only the shape of the matrix Λ changes when we consider different coordinate transformations from the ones presented in (A.21). For example, we can consider the rotations in the planes YOZ and ZOX, or inversions of coordinates, or even more complicated transformations.

Fig. A.1 Transformation of the coordinates of a vector under rotation of φ

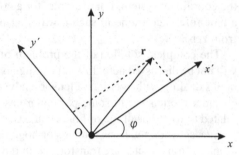

Finally, we will need a general form of matrices Λ and Λ^{-1} in (A.20), which is given by

$$(\Lambda) = \begin{pmatrix} \frac{\partial x}{\partial x'} & \frac{\partial x}{\partial y'} & \frac{\partial x}{\partial z'} \\ \frac{\partial y}{\partial x'} & \frac{\partial y}{\partial y'} & \frac{\partial y}{\partial z'} \\ \frac{\partial z}{\partial x'} & \frac{\partial z}{\partial y'} & \frac{\partial z}{\partial z'} \end{pmatrix} \quad \text{and} \quad (\Lambda^{-1}) = \begin{pmatrix} \frac{\partial x'}{\partial x} & \frac{\partial y'}{\partial x} & \frac{\partial z'}{\partial x} \\ \frac{\partial x'}{\partial y} & \frac{\partial y'}{\partial y} & \frac{\partial z'}{\partial y} \\ \frac{\partial x'}{\partial z} & \frac{\partial y'}{\partial z} & \frac{\partial z'}{\partial z} \end{pmatrix}. \qquad (A.23)$$

Let us note that for the case of rotations these matrices satisfy the conditions

$$(\Lambda)^T = (\Lambda^{-1}), \qquad \det(\Lambda) = 1,$$

while for parity transformations (inversion of all the axes) $\det(\Lambda) = -1$. In fact, here we will not really need these details, which are important in other sections of Physics.

It is easy to check that the transformation described above does not modify the scalar product between two vectors,

$$\mathbf{A} \cdot \mathbf{B} = A_1 B_1 + A_2 B_2 + A_3 B_3. \qquad (A.24)$$

If replacing (A.21) into (A.24), it is easy to verify that the scalar product of the two vectors is a scalar.

A.3 Differential Calculus

The first example of a derivative of a field is the derivative of a scalar field $f(x,y,z)$. It is good to point out that there is no unique definition of such derivative, like in the case of a function of one variable, because the function $f(x,y,z)$ may vary differently in different directions. We can choose three preferred directions and consider the partial derivatives with respect to the variables x, y and z. For instance, in the case of the coordinate x, we define the partial derivative as follows:

$$f'_x = \frac{\partial f}{\partial x} = \lim_{\Delta x \to 0} \frac{f(x+\Delta x, y, z) - f(x, y, z)}{\Delta x}. \tag{A.25}$$

According to the above formula, the partial derivative f'_x is an ordinary derivative with respect to x, when we keep the other two variables, y and z, constants. In other words, it is the derivative in the selected direction x. We can also define the partial derivatives with respect to y and z, according to

$$f'_y = \frac{\partial f}{\partial y} = \lim_{\Delta y \to 0} \frac{f(x, y+\Delta y, z) - f(x, y, z)}{\Delta y},$$

$$f'_z = \frac{\partial f}{\partial z} = \lim_{\Delta z \to 0} \frac{f(x, y, z+\Delta z) - f(x, y, z)}{\Delta z}. \tag{A.26}$$

The next thing we have to learn is to take the derivative of the function $f(\mathbf{r})$ in an arbitrary direction (directional derivative). For example, let us take the derivative in the direction along the unit vector

$$\hat{\mathbf{n}} = \hat{\mathbf{i}}\cos\alpha + \hat{\mathbf{j}}\cos\beta + \hat{\mathbf{k}}\cos\gamma,$$

where $\quad \cos^2\alpha + \cos^2\beta + \cos^2\gamma = 1.$ \hfill (A.27)

To take the derivative in the direction $\hat{\mathbf{n}}$ at the point P, we consider a curve which passes through this point and has $\hat{\mathbf{n}}$ as a tangent vector. Suppose the curve is parameterized by the natural parameter l, that increases in the positive direction of the vector $\hat{\mathbf{n}}$, and satisfies

$$dl^2 = dx^2 + dy^2 + dz^2.$$

We define the derivative in the direction $\hat{\mathbf{n}}$ at the point P as the derivative with respect to the parameter l in this point and calculate this derivative by the chain rule,

$$\frac{df}{dl} = \frac{\partial f}{\partial x}\frac{dx}{dl} + \frac{\partial f}{\partial y}\frac{dy}{dl} + \frac{\partial f}{\partial z}\frac{dz}{dl} = f'_x\cos\alpha + f'_y\cos\beta + f'_z\cos\gamma$$

$$= (f'_x\hat{\mathbf{i}} + f'_y\hat{\mathbf{j}} + f'_z\hat{\mathbf{k}}) \cdot (\hat{\mathbf{i}}\cos\alpha + \hat{\mathbf{j}}\cos\beta + \hat{\mathbf{k}}\cos\gamma) = \operatorname{grad} f \cdot \hat{\mathbf{n}}. \tag{A.28}$$

In the last formula, we introduced an important quantity, the *gradient* of the scalar field,

$$\operatorname{grad} f = \frac{\partial f}{\partial x}\hat{\mathbf{i}} + \frac{\partial f}{\partial y}\hat{\mathbf{j}} + \frac{\partial f}{\partial z}\hat{\mathbf{k}}. \tag{A.29}$$

The vector field grad $f(\mathbf{r})$ is completely defined by the scalar field $f(\mathbf{r})$. To clarify the geometrical sense of the gradient, let us consider different directions of the vector $\hat{\mathbf{n}}$. Of course, the dot product in (A.28) will depend on the direction and this dependence is caused by the different variations of the function f in different directions. Note that, if in a certain direction given by $\hat{\mathbf{n}}$, the variation of the function $f(\mathbf{r})$ is maximal and positive, that means that, according to Eq. (A.28), this $\hat{\mathbf{n}}$ represents the direction of the given gradient. The reason is that the dot product between two vectors with fixed length reaches its maximum value when these two vectors have the same directions.

So, the vector grad $f(\mathbf{r})$ points in the direction of greater growth of the function $f(\mathbf{r})$. Obviously, if we choose the direction of the vector $\hat{\mathbf{n}}$ in the same direction, the derivative df/dl of the Eq. (A.28) must have exactly the same value that the modulus of the gradient. Thus, the magnitude of the gradient vector at the point P is equal to the maximal value of the directional derivative at the same point.

The next step is to examine how the components of the gradient are transformed according to the coordinate transformations. We will see that the gradient of a scalar is an example of vector, this means it follows the law (A.22). For this end, let us consider the coordinate transformation

$$x, y, z \rightarrow x', y', z'.$$

According to the consideration given above, in the general case the transformation has the form of the Eq. (A.20), but with the matrix Λ that can be more general than for the single rotation. Consider now, using the chain rule, the transformation of gradient. For the sake of simplicity, we write only one component,

$$\frac{\partial f}{\partial x} = \frac{\partial f}{\partial x'}\frac{\partial x'}{\partial x} + \frac{\partial f}{\partial y'}\frac{\partial y'}{\partial x} + \frac{\partial f}{\partial z'}\frac{\partial z'}{\partial x}. \tag{A.30}$$

As a simple exercise, the reader can write similar formulas for the other components. The derivatives

$$\frac{\partial x'}{\partial x}, \frac{\partial x'}{\partial y}, \dots, \frac{\partial z'}{\partial z}$$

are components of the matrix Λ^{-1}. The transformation is performed by means of the components of this matrix, so that it fits into the standard formula (A.22).

The next step is to write the gradient in an alternative useful form

$$\text{grad} f = \left(\hat{\mathbf{i}}\frac{\partial}{\partial x} + \hat{\mathbf{j}}\frac{\partial}{\partial y} + \hat{\mathbf{k}}\frac{\partial}{\partial z}\right) f(x,y,z) = \nabla f(\mathbf{r}), \tag{A.31}$$

where we introduced a new object called *nabla operator* or *Hamilton operator*, defined by

$$\nabla = \hat{\mathbf{i}}\frac{\partial}{\partial x} + \hat{\mathbf{j}}\frac{\partial}{\partial y} + \hat{\mathbf{k}}\frac{\partial}{\partial z}. \tag{A.32}$$

When the operator ∇ acts on a scalar field, it produces its gradient. In general, the word "operator" means a mapping of one set of functions to another set of functions. In the case of the gradient operator ∇ makes a mapping of scalars into vectors.

Can we apply ∇ to the vectors? From the algebraic point of view, we can treat the partial derivatives in (A.32) as numerical coefficients and then the operator ∇ is a vector. Thinking in this way, the multiplication of a vector to a scalar is a vector, then it is natural that $\operatorname{grad} f$ is also a vector. Remembering that there are two types of products between two vectors, we can conclude that there are two different possibilities to apply the operator ∇ to the vectors. The first option is similar to the dot product and is called *divergence* of a vector field. For a vector field, $\mathbf{A}$, defined in (A.18), we have

$$\operatorname{div} \mathbf{A} = \nabla \cdot \mathbf{A} = \frac{\partial A_1}{\partial x} + \frac{\partial A_2}{\partial y} + \frac{\partial A_3}{\partial z}. \qquad (A.33)$$

Obviously, the main property is that the divergence of a vector field is a scalar.

The second option is to construct the cross product of ∇ to the vector field $\mathbf{A}$. The result is called *curl* of a vector field and it has the following form:

$$\operatorname{rot} \mathbf{A} = \nabla \times \mathbf{A} = \begin{vmatrix} \hat{\mathbf{i}} & \hat{\mathbf{j}} & \hat{\mathbf{k}} \\ \frac{\partial}{\partial x} & \frac{\partial}{\partial y} & \frac{\partial}{\partial z} \\ A_1 & A_2 & A_3 \end{vmatrix}$$

$$= \hat{\mathbf{i}} \left(\frac{\partial A_3}{\partial y} - \frac{\partial A_2}{\partial z} \right) + \hat{\mathbf{j}} \left(\frac{\partial A_1}{\partial z} - \frac{\partial A_3}{\partial x} \right) + \hat{\mathbf{k}} \left(\frac{\partial A_2}{\partial x} - \frac{\partial A_1}{\partial y} \right). \qquad (A.34)$$

The curl of a vector field is also a vector field. We will consider the geometric interpretation and some properties of the divergence and curl later on, using integral theorems of Gauss (also called Gauss-Ostrogradsky) and Stokes.

When we introduced the gradient, divergence and curl we forgot to check out to which kinds of fields such operations can be applied and to which they cannot. The minimum conditions include the existence of all the partial derivatives at the point in space where these operations are applied. Indeed, for the most interesting situations, from the physical point of view, we can assume that the partial derivatives themselves are continuous functions and also that they are continuously differentiable with respect to their arguments x, y and z. In this case, according to the theorems of Analysis, the second derivatives of these functions do not depend on the order of derivation,

$$\frac{\partial^2 P}{\partial x \partial y} = \frac{\partial^2 P}{\partial y \partial x}, \quad \frac{\partial^2 P}{\partial y \partial z} = \frac{\partial^2 P}{\partial z \partial y}, \quad \frac{\partial^2 P}{\partial z \partial x} = \frac{\partial^2 P}{\partial x \partial z} \qquad (A.35)$$

for any function $P(x, y, z)$. Using these relations, several important properties of gradient, divergence and curl can be found.

Exercises

Demonstrate the following relations:

1. $\operatorname{div}(\operatorname{rot}\mathbf{A}) = 0$;
2. $\operatorname{rot}(\operatorname{grad}f) = 0$;
3. $\operatorname{div}(\operatorname{grad}f) = \Delta f = \frac{\partial^2 f}{\partial x^2} + \frac{\partial^2 f}{\partial y^2} + \frac{\partial^2 f}{\partial z^2}$.
   The symbol Δ is called *Laplace Operator*.
4. $\operatorname{grad}(\varphi\Psi) = \varphi\operatorname{grad}\Psi + \Psi\operatorname{grad}\varphi$;
5. $\operatorname{div}(\varphi\mathbf{A}) = \varphi\operatorname{div}\mathbf{A} + (\mathbf{A}\cdot\operatorname{grad})\varphi = \varphi\nabla\mathbf{A} + (\mathbf{A}\cdot\nabla)\varphi$;
6. $\operatorname{rot}(\varphi\mathbf{A}) = \varphi\operatorname{rot}\mathbf{A} - [\mathbf{A},\nabla]\varphi$;
7. $\operatorname{div}[\mathbf{A},\mathbf{B}] = \mathbf{B}\cdot\operatorname{rot}\mathbf{A} - \mathbf{A}\cdot\operatorname{rot}\mathbf{B}$;
8. $\operatorname{rot}[\mathbf{A},\mathbf{B}] = \mathbf{A}\operatorname{div}\mathbf{B} - \mathbf{B}\operatorname{div}\mathbf{A} + (\mathbf{B},\nabla)\mathbf{A} - (\mathbf{A},\nabla)\mathbf{B}$.
9. $[\mathbf{A},\operatorname{rot}\mathbf{A}] = \frac{1}{2}\operatorname{grad}A^2 - (\mathbf{A},\nabla)\mathbf{A}$.
   Hints. The first three exercises are the most simple ones. The last four have a little bit higher level of difficulty and their solution is much simpler if using the tensor methods.

10. Prove that the divergence of a vector field is a scalar.

A.4 Elements of Integral Calculus for Vectors

In this section we will introduce some necessary elements of the Integral Calculus for vectors. First, we consider integrals along a curved line, also called curvilinear integral, or line integral. One can distinguish two types of curvilinear integrals, integral of the first kind and integral of the second kind. After that we review the integrals over 2D surfaces and over the 3D volumes.

A.4.1 Curvilinear Integral of the First Kind

Consider a simple curve (L) in three dimensional space, 3D. The parametric equation of the curve has the form

$$x = x(\lambda), \quad y = y(\lambda), \quad z = z(\lambda), \tag{A.36}$$

where we have three coordinate functions of the parameter λ. All of them, $x(\lambda), y(\lambda)$ and $z(\lambda)$ are, by assumption, differentiable and their derivatives are continuous functions. The word "simple" with respect to a curve means that the curve does not cross with itself and admits the introduction of the natural parameter l, where

$$dl = \sqrt{dx^2 + dy^2 + dz^2} = \sqrt{x'^2 + y'^2 + z'^2}\,d\lambda.$$

Without loss of generality, we assume that the curve is open and that the natural parameter assumes increasing values with $0 < l < L$ as it moves ahead from the beginning of a curve until the end, while the parameter λ varies, correspondingly, from[1] $\lambda = a$ to $\lambda = b$.

For a continuous function, $f(\mathbf{r}) = f(x, y, z)$, defined along the curve, we define the curvilinear integral of the first kind,

$$I_1 = \int_{(L)} f(\mathbf{r}) \, dl. \qquad (A.37)$$

In order to define this integral, the interval $[a, b]$ should be divided into parts according to

$$a = \lambda_0 < \lambda_1 < \ldots < \lambda_i < \lambda_{i+1} < \ldots < \lambda_n = b, \qquad (A.38)$$

and choose the points $\xi_i \in [\lambda_i, \lambda_{i+1}]$ within each piece, $\Delta\lambda_i = \lambda_{i+1} - \lambda_i$. The length of each partial curve is

$$\Delta l_i = \int_{\lambda_i}^{\lambda_{i+1}} \sqrt{x'^2 + y'^2 + z'^2} \, d\lambda. \qquad (A.39)$$

Now we can construct the integral sum

$$\sigma = \sum_{i=0}^{n-1} f(\mathbf{r}(\xi_i)) \, \Delta l_i. \qquad (A.40)$$

The next step is to consider the limit when

$$\Delta l_{max} = \max\{\Delta l_1, \Delta l_2, \ldots \Delta l_n\} \to 0 \qquad (A.41)$$

when the number of divisions of the original interval tends to infinity. It is possible to show that the expression (A.39) approaches the integral sum σ^* for the Riemann integral,

$$I_1^* = \int_a^b f(x(\lambda), y(\lambda), z(\lambda)) \sqrt{x'^2 + y'^2 + z'^2} \, d\lambda. \qquad (A.42)$$

In the course of Real Analysis, it is shown that the convergence of the integral sum, σ^*, corresponding to (A.42), implies the convergence of the other integral sum, σ, and vice-versa, and their limits under (A.41) are the same. As a result, we come to the representation

---

[1] The parameter λ may be time or any other parameter, including the natural parameter l.

$$I_1 = \int\limits_{(L)} f(\mathbf{r})\, dl = \int\limits_{a}^{b} f(x(\lambda), y(\lambda), z(\lambda))\, \sqrt{x'^2 + y'^2 + z'^2}\, d\lambda = I_1^*. \quad \text{(A.43)}$$

for the curvilinear integral of the first kind. For example, the formula (A.43) can be used to study the properties of the expression (A.37).

The main properties of the curvilinear integral of the first kind are

$$\text{additivity,} \quad \int\limits_{(AB)} f dl + \int\limits_{(BC)} f dl = \int\limits_{(AC)} f dl,$$

$$\text{symmetry,} \quad \int\limits_{(AB)} f dl = \int\limits_{(BA)} f dl.$$

These relationships can be easily obtained by using (A.43). The first relation means that if we separate the curve $(AC) = (L)$ by an intermediate point B, the integral along (AC) will be equal to the sum of the integrals along the parts of the curve. The second relation shows that a change of order of integration does not lead to the change of the sign of the integral. The origin of this feature of the first-kind curvilinear integral is that we use the length of the partial curves, Δl_i, in the construction of the integral sum σ in (A.40). The signs of all Δl_i are positive and therefore it is independent of the direction of integration along the curve.

A.4.2 Curvilinear Integral of Second Kind

The curvilinear integral of the second kind is a scalar that results from the integration of a vector field $\mathbf{F}$ along a curved line (L). Let us consider a vector field $\mathbf{F}$, defined along the curve. In this case, for a displacement $d\mathbf{r}$ between the two infinitesimally close points, we can construct the scalar differential element

$$\mathbf{F} \cdot d\mathbf{r} = F_1 dx + F_2 dy + F_2 dz. \quad \text{(A.44)}$$

If the curve is parameterized by a continuous monotone parameter λ, the last expression can be presented in the form

$$\mathbf{F} \cdot d\mathbf{r} = \left(F_1 \frac{dx}{d\lambda} + F_2 \frac{dy}{d\lambda} + F_3 \frac{dz}{d\lambda} \right) d\lambda, \quad \text{(A.45)}$$

where $F_i = F_i(x(\lambda), y(\lambda), z(\lambda))$ for $i = 1, 2, 3$. The expression (A.45) can be integrated along the curve or, equivalently, we can integrate over the parameter λ. Thus, the integral may be expressed as

$$\int\limits_{(L)} \mathbf{F} \cdot d\mathbf{r} = \int\limits_{(L)} F_1 dx + F_2 dy + F_3 dz \qquad (A.46)$$

$$= \int_a^b \left(F_1 \frac{dx}{d\lambda} + F_2 \frac{dy}{d\lambda} + F_3 \frac{dz}{d\lambda} \right) d\lambda. \qquad (A.47)$$

In the expression (A.47), we can see how to reduce the curvilinear integral of the second kind (A.46) to a common definite integral of Riemann.

Each of the three integrals in (A.46) can be defined as the limit of the integral sum for the corresponding integral. Let us consider this reduction for the case of one of these integrals and leave the other two as an exercise for the reader. We have

$$\int\limits_{(L)} F_1 dx = \lim_{\Delta l_{max} \to 0} \sigma_x \qquad (A.48)$$

where

$$\sigma_x = \sum_{i=0}^{n-1} F_1\left(\mathbf{r}(\xi_i)\right) \cdot \Delta x_i. \qquad (A.49)$$

In this formula, we introduced a new notation, compared to the Eq. (A.38), namely

$$\Delta x_i = x(\lambda_{i+1}) - x(\lambda_i).$$

The limit of the integral sum should be taken exactly in the same manner as in the previous case of the integral of the first kind. We leave it to the reader to look for more details in books of Analysis.

Finally, we can rewrite the representation (A.47) using cosines defined in (A.27),

$$\cos\alpha = \frac{d\mathbf{r} \cdot \hat{\mathbf{i}}}{dl} = \frac{dx}{dl},$$

$$\cos\beta = \frac{d\mathbf{r} \cdot \hat{\mathbf{j}}}{dl} = \frac{dy}{dl},$$

$$\cos\gamma = \frac{d\mathbf{r} \cdot \hat{\mathbf{k}}}{dl} = \frac{dz}{dl}. \qquad (A.50)$$

By using these notations we arrive at

$$\int\limits_{(L)} \mathbf{F} \cdot d\mathbf{r} = \int_0^L \left(F_x \cos\alpha + F_y \cos\beta + F_z \cos\gamma \right) dl$$

$$= \int_a^b \left(F_x \cos\alpha + F_y \cos\beta + F_z \cos\gamma \right) \sqrt{\dot{x}^2 + \dot{y}^2 + \dot{z}^2}\, d\lambda, \qquad (A.51)$$

where $\dot{x} = dx/d\lambda$ and the same for the other two coordinates.

The main properties of the integral of second kind are the following:

- additivity $$\int\limits_{AB} \mathbf{F} \cdot d\mathbf{r} + \int\limits_{BC} \mathbf{F} \cdot d\mathbf{r} = \int\limits_{AC} \mathbf{F} \cdot d\mathbf{r};$$

- antisymmetry $$\int\limits_{AB} \mathbf{F} \cdot d\mathbf{r} = - \int\limits_{BA} \mathbf{F} \cdot d\mathbf{r}. \qquad (A.52)$$

The first property is common also to the usual Riemann integral and can be verified by analyzing the integral sum or through the representation (A.47) or (A.51). The second property also comes from the one of the common definite integral. It differs from the case of the curvilinear integral of the first kind. For example, for the representation (A.51), this property can be checked if we observe that the unit tangent vector, $\hat{\mathbf{n}}$, reverses its direction when we change the direction of integration. As a result, all three cosines change their signs.

Finally, consider a simple closed curve, (C) (also called contour), and choose one direction on this curve as positive. In this way we obtain an oriented contour, let us call it (C^+). The integral of a field $\mathbf{A}$ along the positively oriented curve is called the circulation of the vector field $\mathbf{A}$ over the curve (C^+). It can be expressed as

$$\oint\limits_{(C^+)} \mathbf{A} \cdot d\mathbf{r}. \qquad (A.53)$$

Obviously, the circulation changes its sign when we invert the orientation of the contour.

A.5 Surface and Volume Integrals in 3D Space

Consider the integrals over the $2D$ surface (S) in the $3D$ space. Suppose the surface is defined by three smooth functions

$$x = x(u,v), \qquad y = y(u,v), \qquad z = z(u,v), \qquad (A.54)$$

where u and v are called internal coordinates on the surface. One can regard (A.54) as a mapping of a figure (G) on the plane with the coordinates u, v into the $3D$ space with the coordinates x, y, z, as illustrated in Fig. A.2.

Example 2. The sphere (S) of constant radius R can be described by internal (angular) coordinates φ, θ. Then the relation (A.54) is cast into the form

$$x = R\cos\varphi\sin\theta, \qquad y = R\sin\varphi\sin\theta, \qquad z = R\cos\theta, \qquad (A.55)$$

where φ and θ play the role of u and v. The values of these coordinates are taken from the rectangle

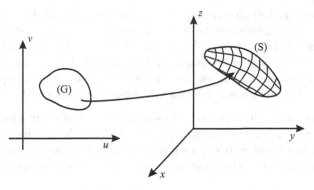

Fig. A.2 Picture of a mapping of (G) into (S)

$$(G) = \{(\varphi, \theta) \mid 0 \leq \varphi < 2\pi, 0 \leq \theta < \pi\},$$

and the parametric description of the sphere (A.55) is nothing but the mapping $(G) \rightarrow (S)$. One can express all geometric quantities on the surface of the sphere using the internal coordinates φ and θ, without addressing the coordinates of the external space x, y, z.

The geometry of the surfaces is discussed in many text-books on Calculus and Differential Geometry, one can see, for example, [29]. The only one result we will need is the expression for the differential element of the surface area, which has the form

$$dS = \sqrt{g}\, du\, dv, \tag{A.56}$$

where

$$g = \begin{vmatrix} g_{uu} & g_{uv} \\ g_{vu} & g_{vv} \end{vmatrix}$$

is the so called metric determinant. The components of the metric (the complete name is "induced metric tensor on the surface") are given by the expressions

$$\begin{aligned}
g_{uu} &= x_u'^2 + y_u'^2 + z_u'^2, \\
g_{uv} &= x_u' x_v' + y_u' y_v' + z_u' z_v', \\
g_{vv} &= x_v'^2 + y_v'^2 + z_v'^2.
\end{aligned} \tag{A.57}$$

Using (A.56), the area S of the whole surface (S) may be presented in a form of a double integral over (G),

$$S = \iint_{(G)} \sqrt{g}\, du\, dv. \tag{A.58}$$

Geometrically, this definition corresponds to a triangulation of the curved surface with the consequent limit, when the size of all triangles simultaneously tend to zero. The coefficient $\sqrt{g}$ in (A.56) and (A.58) plays the role of the Jacobian of the transformation of coordinates in $2D$, but there is an essential difference, because the matrix of transformation from internal coordinates u and v is not a square matrix and, therefore, the determinant of such a transformation cannot be defined.

Now we are in a position to define the surface integral of the first kind. Suppose we have a surface (S) defined as a mapping (A.54) of the closed finite figure (G) in a uv-plane, and a function $f(u,v)$ on this surface. Instead one can take a function $g(x,y,z)$ which generates a function on the surface through the relation

$$f(u,v) = g\big(x(u,v),\, y(u,v),\, z(u,v)\big).\tag{A.59}$$

The surface integral of the first kind is defined as follows: Let us divide (S) into particular sub-surfaces (S_i) such that each of them has a well-defined area

$$S_i = \iint\limits_{G_i} \sqrt{g}\,du\,dv.\tag{A.60}$$

Of course, one has to perform the division of (S) into (S_i) such that the intersections of the two sub-figures have zero area. On each of the particular surfaces (S_i) we choose a point $M_i(\xi_i, \eta_i)$ and construct an integral sum

$$\sigma = \sum_{i=1}^{n} f(M_i)\cdot S_i,\tag{A.61}$$

where $f(M_i) = f(\xi_i, \eta_i)$. The next step is to define $\lambda = \max\{\operatorname{diam}(S_i)\}$.[2] If the limit

$$I_1 = \lim_{\lambda \to 0} \sigma\tag{A.62}$$

is finite and does not depend on the choice of (S_i) and on the choice of M_i, it is called the surface integral of the first kind

$$I_1 = \iint\limits_{(S)} f(u,v)\,dS = \iint\limits_{(S)} g(x,y,z)\,dS.\tag{A.63}$$

It is clear from the last relation that this integral can be indeed calculated as a double integral over the figure (G) in the uv-plane

$$I_1 = \iint\limits_{(S)} f(u,v)\,dS = \iint\limits_{(G)} f(u,v)\,\sqrt{g}\,du\,dv.\tag{A.64}$$

---

[2] The diameter $\operatorname{diam}(S)$ is the greatest distance between the two points of the closed set.

Observation. The surface integral (A.64) is, by construction, a scalar, if the function $f(u,v)$ is a scalar. However it is not forbidden to take such an integral from the vector field too, the result will be a constant vector.

Let us now consider a more complicated type of integral, namely the surface integral of the second kind. The relation between the two types of surface integrals is similar to that between the two types of curvilinear integrals, which we have considered in the previous section.

First of all, one has to introduce the notion of a smooth oriented surface. We assume that the surface has two distinct sides. This means that if we take a normal unit vector to the surface, $\hat{\mathbf{n}}$, and start moving it to other points of the surface, (it should always be a normal unit vector), the direction of $\hat{\mathbf{n}}$ will never change to the opposite. Let us note that there are examples, e.g., the Mobius surface, for which this condition is not satisfied.

For calculational purposes it is useful to introduce the components of $\hat{\mathbf{n}}$, using the same cosines which we already considered for the curvilinear second kind integral,

$$\cos\alpha = \hat{\mathbf{n}}\cdot\hat{\mathbf{i}}, \qquad \cos\beta = \hat{\mathbf{n}}\cdot\hat{\mathbf{j}}, \qquad \cos\gamma = \hat{\mathbf{n}}\cdot\hat{\mathbf{k}}. \qquad (A.65)$$

Now we are able to formulate the definition of the second-kind surface integral. Consider a continuous vector field $\mathbf{A}(\mathbf{r})$ defined on the surface (S). The surface integral of the second kind is

$$\iint\limits_{(S)} \mathbf{A} \cdot d\mathbf{S} = \iint\limits_{(S)} \mathbf{A} \cdot \mathbf{n}\, dS. \qquad (A.66)$$

It is the universal (if it exists, of course) limit of the integral sum

$$\sigma = \sum_{i=1}^{n} \mathbf{A}_i \cdot \hat{\mathbf{n}}_i S_i, \qquad (A.67)$$

where both vectors must be taken in the same point $M_i \in (S_i)$: $\mathbf{A}_i = \mathbf{A}(M_i)$ and $\hat{\mathbf{n}}_i = \hat{\mathbf{n}}(M_i)$. Other notations here are the same as in the case of the surface integral of the first kind, that is

$$(S) = \bigcup_{i=1}^{n} (S_i), \qquad \lambda = \max\{\operatorname{diam}(S_i)\}$$

and the integral corresponds to the limit $\lambda \to 0$.

By construction, the surface integral of the second kind equals to the following double integral over the figure (G) in the uv-plane:

$$\iint\limits_{(S)} \mathbf{A} \cdot d\mathbf{S} = \iint\limits_{(G)} (A_x \cos\alpha + A_y \cos\beta + A_z \cos\gamma)\, \sqrt{g}\, du dv. \qquad (A.68)$$

Observation. One can prove that the surface integral exists if $\mathbf{A}(u,v)$ and $\mathbf{r}(u,v)$ are smooth functions and the surface (S) has finite area. The sign of the integral changes if we change the orientation of the surface to the opposite one $\hat{\mathbf{n}} \to -\hat{\mathbf{n}}$.

There are important relations between surface integral of the second kind, curvilinear integral of the second kind and the $3D$ volume integral. These relations are called theorems of Stokes and Gauss. Let us formulate these theorems without proof.

Theorem 1 (Stokes' Theorem). *For any continuous vector function* $\mathbf{F}(\mathbf{r})$ *with continuous partial derivatives, the following relation holds:*

$$\iint\limits_{(S)} rot\mathbf{F} \cdot d\mathbf{S} = \oint\limits_{(\partial S^+)} \mathbf{F} \cdot d\mathbf{r}. \tag{A.69}$$

The last integral in (A.69) is a special case of curvilinear integral of the second kind, which is taken over the closed contour. It is called a circulation of the vector field $\mathbf{F}$ over the closed oriented path (∂S^+). The orientation of the path depends on the direction of $\hat{\mathbf{n}}$ in the vector element of the area, $d\mathbf{S} = \hat{\mathbf{n}}dS$. The choice of (∂S^+) means that the motion along the path is performed clockwise if we look at the positive direction of $\hat{\mathbf{n}}$.

An important consequence of the Stokes Theorem is that for the $\mathbf{F}(\mathbf{r}) = \operatorname{grad} U(\mathbf{r})$, where $U(\mathbf{r})$ is some smooth function, the curvilinear integral between two points A and B doesn't depend on the path (AB) and is simply

$$\int\limits_{(AB)} \mathbf{F} \cdot d\mathbf{r} = U(B) - U(A).$$

We leave the proof of this statement as an exercise for an interested reader.

The last Theorem we are going to formulate is the

Theorem 2 (Gauss-Ostrogradsky Theorem). *Consider a 3D figure* $(V) \subset R^3$ *and also define* (∂V^+) *to be the externally oriented boundary of* (V). *Consider a vector field* $\mathbf{E}(\mathbf{r})$ *defined on* (V) *and on its boundary* (∂V^+) *and suppose that the components of this vector field are continuous, as are their partial derivatives*

$$\frac{\partial E^x}{\partial x}, \qquad \frac{\partial E^y}{\partial y}, \qquad \frac{\partial E^z}{\partial z}.$$

Then these components satisfy the following integral relation

$$\oiint\limits_{(\partial V^+)} \mathbf{E} \cdot d\mathbf{S} = \iiint\limits_{(V)} div\mathbf{E}\, dV. \tag{A.70}$$

Let us finally note that relatively simple geometric proofs of both Stokes's and Gauss theorems one can find, e.g., in [29] and, of course, in many other text-books.

Using the Stokes and Gauss-Ostrogradsky theorems, one can provide a more explicit geometric interpretation of divergence and curl of the vector. Suppose we want to know the projection of $\text{rot}\,\mathbf{F}$ on the direction of the unit vector $\hat{\mathbf{n}}$. Then we have to take some infinitesimal surface vector $d\mathbf{S} = dS\,\hat{\mathbf{n}}$. In this case, due to the continuity of all components of the vector $\text{rot}\,\mathbf{F}$, we have

$$\hat{\mathbf{n}} \cdot \text{rot}\,\mathbf{F} = \lim_{dS \to 0} \frac{1}{dS} \oint_{(\partial S+)} \mathbf{F} \cdot d\mathbf{r}. \tag{A.71}$$

where (∂S^+) is a border of (S). The last relation (A.71) provides a necessary food for the geometric intuition of what $\text{rot}\,\mathbf{F}$ is.

In the case of $\text{div}\,\mathbf{A}$ one meets the following relation

$$\text{div}\,\mathbf{A} = \lim_{V \to 0} \frac{1}{V} \oiint_{(\partial V+)} \mathbf{A} \cdot d\mathbf{S}, \tag{A.72}$$

where the limit $V \to 0$ must be taken in such a way that also $\lambda \to 0$, where $\lambda = \text{diam}\,(V)$. This formula indicates the relation between $\text{div}\,\mathbf{A}$ at some point and the presence of the source for this field at this point. If such source is absent, $\text{div}\,\mathbf{A} = 0$.

The two formulas (A.72) and (A.71) make the geometric sense of div and rot operators sufficiently explicit.

References

1. R. Abraham, J.E. Marsden, *Foundations of Mechanics* (AMS Chelsea, Providence, 2008)
2. G.B. Arfken, H.J. Weber, *Mathematical Methods for Physicists* (Academic, San Diego, 1995)
3. V.I. Arnold, *Mathematical Methods of Classical Mechanics* (Springer, New York, 1989)
4. V.I. Arnold, *Ordinary Differential Equations*, 3rd edn. (Springer, 1997)
5. J. Barcelos Neto, *Mecânica Newtoniana, Lagrangiana & Hamiltoniana* (Editora Livraria da Física, São Paulo, 2004)
6. G.K. Batchelor, *An Introduction to Fluid Dynamics* (Cambridge Mathematical Library, Cambridge University Press,Cambridge/New York, 2000)
7. M. Braun, *Differential Equations and Their Applications: An Introduction to Applied Mathematics*. Texts in Applied Mathematics, vol. 11, 4th edn. (Springer, New York, 1992)
8. H.C. Corben, P. Stehle, *Classical Mechanics* (Dover, New York, 1994)
9. E.A. Desloge, *Classical Mechanics*, vols. 1 and 2 (Wiley, New York, 1982)
10. F.R. Gantmacher, *Lectures in Analytical Mechanics* (Mir Publications, Moscow, 1975)
11. H. Goldstein, C.P. Poole, J.L. Safko, *Classical Mechanics*, 3rd edn. (Addison Wesley, San Francisco, 2001)
12. I.S. Gradshteyn, I.M. Ryzhik, *Table of Integrals, Series, and Products*, 7th edn. (Elsevier, Burlington, 2007)
13. R.A. Granger, *Fluid Mechanics*. Dover Books on Physics (Dover, New York, 1995)
14. D. Gregory, *Classical Mechanics* (Cambridge University Press, Cambridge, 2006)
15. W. Greiner, *Classical Mechanics: Point Particles and Relativity*. Classical Theoretical Physics (Springer, New York/London, 2003)
16. I.E. Irodov, *Mechanics. Basic Laws.* (in Russian) (Nauka, 6 Ed., 2002); I.E. Irodov, *Problems in General Physics* (Mir Publications, 1988 and GK Publisher, 2008)
17. J.D. Jackson, *Classical Electrodynamics*, 3rd edn. (Wiley, New York, 1998)
18. J.V. José, E.J. Saletan, *Classical Dynamics: A Contemporary Approach* (Cambridge University Press, Cambridge/New York, 1998)
19. T.W.B. Kibble, F.H. Berkshire, *Classical Mechanics* (World Scientific, River Edge, 2004)
20. C. Lanczos, *The Variational Principles of Mechanics* (Dover, New York, 1970)
21. L.D. Landau, E.M. Lifshits, *Course of Theoretical Physics: Mechanics*, 3rd edn. (Butterworth-Heinemann, 1982)
22. L.D. Landau, E.M. Lifshitz, *Hydrodynamics*. Course of Theoretical Physics Series, vol. 6, 2nd edn. (Butterworth-Heinemann, 1987)
23. L.D. Landau, E.M. Lifshitz, *The Classical Theory of Fields*. Course of Theoretical Physics Series, vol. 2
24. N.A. Lemos, *Mecânica Analítica*, 2nd edn. (Livaria da Física, 2007)
25. Y.-K. Lim, *Problems and Solutions on Mechanics: Major American Universities Ph.D. Qualifying Questions and Solutions* (World Scientific, Singapore/River Edge, 1994)

I.L. Shapiro and G. de Berredo-Peixoto, *Lecture Notes on Newtonian Mechanics*,   247
Undergraduate Lecture Notes in Physics, DOI 10.1007/978-1-4614-7825-6,
© Springer Science+Business Media, LLC 2013

26. J.B. Marion, S.T. Thornton, *Classical Dynamics of Particle and Systems* (Harcourt, Fort Worth, 1995)
27. D. Morin, *Introduction to Classical Mechanics with Problems and Solutions* (Cambridge University Press, Cambridge/New York, 2008)
28. P.M. Morse, H. Fishbach, *Methods of Theoretical Physics* (McGraw-Hill, New York, 1953)
29. I.L. Shapiro, *Lecture Notes on Vector and Tensor Algebra and Analysis* (CBPF, Rio de Janeiro, 2003)
30. A. Sommerfeld, *Mechanics*. Lectures on Theoretical Physics, vol. 1 (Academic, New York, 1970)
31. K. Symon, *Mechanics*, 3rd edn. (Addison Wesley, Reading, 1971)
32. J.R. Taylor, *Classical Mechanics* (University Science Books, Sausalito, 2005)

Index

I.L. Shapiro and G. de Berredo-Peixoto, *Lecture Notes on Newtonian Mechanics*,
Undergraduate Lecture Notes in Physics, DOI 10.1007/978-1-4614-7825-6,
© Springer Science+Business Media, LLC 2013